TERNARY QUADRATIC FORMS AND NORMS

PURE AND APPLIED MATHEMATICS

A Program of Monographs, Textbooks, and Lecture Notes

Contributions to *Lecture Notes in Pure and Applied Mathematics* are reproduced by direct photography of the author's typewritten manuscript. Potential authors are advised to submit preliminary manuscripts for review purposes. After acceptance, the author is responsible for preparing the final manuscript in camera-ready form, suitable for direct reproduction. Marcel Dekker, Inc. will furnish instructions to authors and special typing paper. Sample pages are reviewed and returned with our suggestions to assure quality control and the most attractive rendering of your manuscript. The publisher will also be happy to supervise and assist in all stages of the preparation of your camera-ready manuscript.

LECTURE NOTES
IN PURE AND APPLIED MATHEMATICS

1. *N. Jacobson*, Exceptional Lie Algebras
2. *L.-Å. Lindahl and F. Poulsen*, Thin Sets in Harmonic Analysis
3. *I. Satake*, Classification Theory of Semi-Simple Algebraic Groups
4. *F. Hirzebruch, W. D. Newmann, and S. S. Koh*, Differentiable Manifolds and Quadratic Forms (out of print)
5. *I. Chavel*, Riemannian Symmetric Spaces of Rank One (out of print)
6. *R. B. Burckel*, Characterization of C(X) Among Its Subalgebras
7. *B. R. McDonald, A. R. Magid, and K. C. Smith*, Ring Theory: Proceedings of the Oklahoma Conference
8. *Y.-T. Siu*, Techniques of Extension of Analytic Objects
9. *S. R. Caradus, W. E. Pfaffenberger, and B. Yood*, Calkin Algebras and Algebras of Operators on Banach Spaces
10. *E. O. Roxin, P.-T. Liu, and R. L. Sternberg*, Differential Games and Control Theory
11. *M. Orzech and C. Small*, The Brauer Group of Commutative Rings
12. *S. Thomeier*, Topology and Its Applications
13. *J. M. López and K. A. Ross*, Sidon Sets
14. *W. W. Comfort and S. Negrepontis*, Continuous Pseudometrics
15. *K. McKennon and J. M. Robertson*, Locally Convex Spaces
16. *M. Carmeli and S. Malin*, Representations of the Rotation and Lorentz Groups: An Introduction
17. *G. B. Seligman*, Rational Methods in Lie Algebras
18. *D. G. de Figueiredo*, Functional Analysis: Proceedings of the Brazilian Mathematical Society Symposium
19. *L. Cesari, R. Kannan, and J. D. Schuur*, Nonlinear Functional Analysis and Differential Equations: Proceedings of the Michigan State University Conference
20. *J. J. Schäffer*, Geometry of Spheres in Normed Spaces
21. *K. Yano and M. Kon*, Anti-Invariant Submanifolds
22. *W. V. Vasconcelos*, The Rings of Dimension Two
23. *R. E. Chandler*, Hausdorff Compactifications
24. *S. P. Franklin and B. V. S. Thomas*, Topology: Proceedings of the Memphis State University Conference
25. *S. K. Jain*, Ring Theory: Proceedings of the Ohio University Conference
26. *B. R. McDonald and R. A. Morris*, Ring Theory II: Proceedings of the Second Oklahoma Conference
27. *R. B. Mura and A. Rhemtulla*, Orderable Groups
28. *J. R. Graef*, Stability of Dynamical Systems: Theory and Applications
29. *H.-C. Wang*, Homogeneous Banach Algebras
30. *E. O. Roxin, P.-T. Liu, and R. L. Sternberg*, Differential Games and Control Theory II
31. *R. D. Porter*, Introduction to Fibre Bundles
32. *M. Altman*, Contractors and Contractor Directions Theory and Applications
33. *J. S. Golan*, Decomposition and Dimension in Module Categories
34. *G. Fairweather*, Finite Element Galerkin Methods for Differential Equations
35. *J. D. Sally*, Numbers of Generators of Ideals in Local Rings
36. *S. S. Miller*, Complex Analysis: Proceedings of the S.U.N.Y. Brockport Conference
37. *R. Gordon*, Representation Theory of Algebras: Proceedings of the Philadelphia Conference
38. *M. Goto and F. D. Grosshans*, Semisimple Lie Algebras
39. *A. I. Arruda, N. C. A. da Costa, and R. Chuaqui*, Mathematical Logic: Proceedings of the First Brazilian Conference

40. *F. Van Oystaeyen*, Ring Theory: Proceedings of the 1977 Antwerp Conference
41. *F. Van Oystaeyen and A. Verschoren*, Reflectors and Localization: Application to Sheaf Theory
42. *M. Satyanarayana*, Positively Ordered Semigroups
43. *D. L. Russell*, Mathematics of Finite-Dimensional Control Systems
44. *P.-T. Liu and E. Roxin*, Differential Games and Control Theory III: Proceedings of the Third Kingston Conference, Part A
45. *A. Geramita and J. Seberry*, Orthogonal Designs: Quadratic Forms and Hadamard Matrices
46. *J. Cigler, V. Losert, and P. Michor*, Banach Modules and Functors on Categories of Banach Spaces
47. *P.-T. Liu and J. G. Sutinen*, Control Theory in Mathematical Economics: Proceedings of the Third Kingston Conference, Part B
48. *C. Byrnes*, Partial Differential Equations and Geometry
49. *G. Klambauer*, Problems and Propositions in Analysis
50. *J. Knopfmacher*, Analytic Arithmetic of Algebraic Function Fields
51. *F. Van Oystaeyen*, Ring Theory: Proceedings of the 1978 Antwerp Conference
52. *B. Kedem*, Binary Time Series
53. *J. Barros-Neto and R. A. Artino*, Hypoelliptic Boundary-Value Problems
54. *R. L. Sternberg, A. J. Kalinowski, and J. S. Papadakis*, Nonlinear Partial Differential Equations in Engineering and Applied Science
55. *B. R. McDonald*, Ring Theory and Algebra III: Proceedings of The Third Oklahoma Conference
56. *J. S. Golan*, Structure Sheaves over a Noncommutative Ring
57. *T. V. Narayana, J. G. Williams, and R. M. Mathsen*, Combinatorics, Representation Theory and Statistical Methods in Groups: YOUNG DAY Proceedings
58. *T. A. Burton*, Modeling and Differential Equations in Biology
59. *K. H. Kim and F. W. Roush*, Introduction to Mathematical Consensus Theory
60. *J. Banas and K. Goebel*, Measures of Noncompactness in Banach Spaces
61. *O. A. Nielsen*, Direct Integral Theory
62. *J. E. Smith, G. O. Kenny, and R. N. Ball*, Ordered Groups: Proceedings of the Boise State Conference
63. *J. Cronin*, Mathematics of Cell Electrophysiology
64. *J. W. Brewer*, Power Series Over Commutative Rings
65. *P. K. Kamthan and M. Gupta*, Sequence Spaces and Series
66. *T. G. McLaughlin*, Regressive Sets and the Theory of Isols
67. *T. L. Herdman, S. M. Rankin, III, and H. W. Stech*, Integral and Functional Differential Equations
68. *R. Draper*, Commutative Algebra: Analytic Methods
69. *W. G. McKay and J. Patera*, Tables of Dimensions, Indices, and Branching Rules for Representations of Simple Lie Algebras
70. *R. L. Devaney and Z. H. Nitecki*, Classical Mechanics and Dynamical Systems
71. *J. Van Geel*, Places and Valuations in Noncommutative Ring Theory
72. *C. Faith*, Injective Modules and Injective Quotient Rings
73. *A. Fiacco*, Mathematical Programming with Data Perturbations I
74. *P. Schultz, C. Praeger, and R. Sullivan*, Algebraic Structures and Applications: Proceedings of the First Western Australian Conference on Algebra
75. *L. Bican, T. Kepka, and P. Němec*, Rings, Modules, and Preradicals
76. *D. C. Kay and M. Breen*, Convexity and Related Combinatorial Geometry: Proceedings of the Second University of Oklahoma Conference

77. *P. Fletcher and W. F. Lindgren*, Quasi-Uniform Spaces
78. *C.-C. Yang*, Factorization Theory of Meromorphic Functions
79. *O. Taussky*, Ternary Quadratic Forms and Norms

Other Volumes in Preparation

TERNARY QUADRATIC FORMS AND NORMS

edited by
Olga Taussky

Department of Mathematics
California Institute of Technology
Pasadena, California

MARCEL DEKKER, INC. New York and Basel

Library of Congress Cataloging in Publication Data

Taussky, Olga.
Ternary quadratic forms and norms.

(Lecture notes in pure and applied mathematics ; v. 79)
1. Forms, Quadratic. I. Title. II. Series.
QA243.T35 1982 512.9'44 82-10021
ISBN 0-8247-1651-5 AACR2

MARCEL DEKKER, INC.
270 Madison Avenue, New York, New York 10016

Current printing (last digit):
10 9 8 7 6 5 4 3 2 1

PRINTED IN THE UNITED STATES OF AMERICA

PREFACE

These notes concern some of the lectures given in a number-theory seminar at the California Institute of Technology in 1974/75 when a particularly large audience was present in connection with Professor Hans Zassenhaus' stay there as Fairchild Distinguished Scholar.

The lectures included here present either a more detailed or a different version of published results.

The editor's interest in the subject was stimulated by two areas of her recent research in which ternary forms representing zero in a nontrivial way entered.

Professor B. L. van der Waerden kindly gave permission for the inclusion of the translation by Zassenhaus of two letters connected with the subject. H. Zassenhaus consented to report on Gauss' work on quadratic forms, in particular ternary forms and n-ary forms representing zero. He incorporated his own approaches and results.

The Notes were issued in preprint form in 1976. In the meantime there has been further development in some of the subjects treated in the Notes. Zassenhaus extended some of Taussky's work in an article, "A theorem on cyclic algebras," in Number Theory and Algebra, Academic Press (1977), 363-393. Estes studied related problems for abstract rings instead of the ring of rational integers in "Determinants of Galois automorphisms of maximal commutative rings of 2×2 matrices" in Linear Algebra and Applications *27* (1979), 225-243. The squaring of the quadratic form mentioned in Taussky's and van der Waerden's articles has now been generalized by her to composition of binary integral quadratic forms with the same discriminant. The quaternion treatment by Venkow of Gauss' result on sums of three squares, treated by H. P. Rehm, has led to the thesis of P. Hanlon, California Institute of Technology (1981), and further work by P. Hanlon and P. Morton. Rehm, linking Venkow's ideas to Taussky's work on ideal matrices, prepared a manuscript entitled "Eine Permutationsdarstellung von Idealklassengruppen."

Olga Taussky

CONTENTS

CONTRIBUTORS

HERSHY KISILEVSKY,* Department of Mathematics, California Institute of Technology, Pasadena, California

WILHELM PLESKEN,† Department of Mathematics, California Institute of Technology, Pasadena, California

HANS PETER REHM,‡ Department of Mathematics, California Institute of Technology, Pasadena, California

OLGA TAUSSKY, Department of Mathematics, California Institute of Technology, Pasadena, California

B. L. van der WAERDEN, Department of Mathematics, University of Zürich, Zürich, Switzerland

ALFRED WEISS, Department of Mathematics, University of Alberta, Edmonton, Canada

HANS J. ZASSENHAUS, Department of Mathematics, Ohio State University, Columbus, Ohio

---

*Current affiliation: Department of Mathematics, Concordia University, Montreal, Canada

†Current affiliation: Lehrstuhl D für Mathematik, Rheinisch-Westfälische Techniche Hochschule, Aachen, Federal Republic of Germany

‡Current affiliation: Mathematisches Institut II, Universitat KARLSRUHE, West Germany

TERNARY QUADRATIC FORMS AND NORMS

THE RÉDEI-REICHARDT THEOREM: ANOTHER PROOF

Hershy Kisilevsky*

Department of Mathematics
California Institute of Technology
Pasadena, California

Let $F = \mathbb{Q}(\sqrt{D})$ where D is the discriminant of the field F. Denote by C, the Sylow 2-subgroup of the strict (i.e., narrow) ideal class group of F. If $G = g(F/\mathbb{Q})$ is the Galois group of $F/\mathbb{Q}$, and σ is a generating automorphism of G, then there is a natural action of G on C. Further, if $x \in C$ and A is an ideal of F representing x, then σx is represented by σA. Since $A \cdot \alpha A = (N_{F/\mathbb{Q}}(A))$ is a principal ideal generated by a rational number, we see that $\sigma x = x^{-1}$ for all $x \in C$. In particular, $\sigma x = x$ if and only if $x^2 = 1$, so it follows that $C^G = C_2$, where $C^G = \{x | \sigma x = x\}$ and $C_2 = \{x | x^2 = 1\}$.

LEMMA. Every element of C^G is represented by an invariant ideal. Hence every element of order 2 in C is represented by a product of ramified primes.

Proof. Let $x \in C$, $\alpha x = x$. If A represents x, then $\alpha A = A(\alpha)$ where $\alpha \in F$ has a positive norm. Taking norms, we see that $N_{F/\mathbb{Q}}(\alpha) = 1$, so by Hilbert's theorem 90, $\alpha = \beta/\sigma\beta$. Hence $\alpha(A(\beta)) = A(\beta)$. Furthermore, $\beta\sigma\beta = \alpha(\sigma\beta)^2$, so if F is real, α can be chosen positive and hence $A(\beta)$ also represents x and is invariant under G. Since the ideals of F invariant under G are products of ramified primes and rational primes, we see that every class of order 2 in C can be represented by a product of ramified primes. □

---

*Current affiliation: Department of Mathematics, Concordia University, Montreal, Canada.

We wish to compute the number k of invariants of C which are divisible by 4, i.e., if the structure of C as an abelian group is given by $(2, \ldots, 2, 2^{a_1}, \ldots, 2^{a_k})$ where $1 < a_1 \leq \cdots \leq a_k$, then the number of invariants divisible by 4 is k.

For an abelian 2-group A of type $(2, \ldots, 2, 2^{a_1}, \ldots, 2^{a_k})$ as above, (A/A_2) has type $(2^{a_1-1}, \ldots, 2^{a_k-1})$ so that k = the number of invariants divisible by 4 = 2-rank of A/A_2. Further, if B is an abelian 2-group of 2-rank t, then

$$2^t = |B_2| = |\{x \in B | x^2 = 1\}|$$
$$= \text{number of subgroups of index 2 in B}$$

Hence letting $B = A/A_2$, we see that 2^k = number of subgroups of index 2 in A/A_2 = the number of subgroups of index 2 in A which contain A_2. In other words, to count the number of invariants of A divisible by 4, it is sufficient to count the number of subgroups H, $A \supseteq H \supseteq A_2$, such that $[A:H] = 2$.

We apply this to the class group C, so that 2^k = the number of subgroups H of C, $C \supseteq H \supseteq C^G$, such that $[C:H] = 2$.

To this end we apply some results of class field theory.

Let K be the class field to C, so that K/F is the maximal abelian extension of F such that [K:F] is a power of 2 and K/F is unramified at all finite primes.

By class field theory, $g(K|F) \approx C$ as G-modules, and since $\sigma \in G$ acts on C by inversion, it follows that every subgroup of $g(K/F)$ is also a G-module, and hence all subfields k, $F \subseteq k \subseteq K$ are in fact normal over $\mathbb{Q}$.

Identify $g(K/F)$ with C. We construct a one-to-one correspondence between subgroups H, $C \supseteq H \supseteq C^G$, $[C:H] = 2$, and D-factorizations of the second kind.

(*Note:* If we omit the condition that $H \supseteq C^G$, we obtain the Gauss theorem for the 2-rank of C.)

Let $C \supseteq H \supseteq C^G$, $[C:H] = 2$. Let $k_2 \supseteq K$ be the subfield of K fixed by H, so that $\mathbb{Q} \subseteq F \subseteq k_2$ and $[k_2:\mathbb{Q}] = 4$. Since $k_2/\mathbb{Q}$ is normal, $g(k_2/\mathbb{Q})$ is either cyclic or $\mathbb{Z}/2 \times \mathbb{Z}/2$.

Suppose $g(k_2/\mathbb{Q}) = \mathbb{Z}/4$. Then $g(k_2/F)$ is the unique subgroup of $g(k_2/\mathbb{Q})$ of order 2. For any rational prime p, the ramification index of p in k_2 is at most 2 as k_2/F is unramified, and so the inertial group of p has order at most 2 and thus must be a subgroup of $g(k_2/F)$. But this would mean that $F/\mathbb{Q}$ is unramified, which is a contradiction. Thus $g(k_2/\mathbb{Q}) = \mathbb{Z}/2 \times \mathbb{Z}/2$ and $k_2 = \mathbb{Q}(\sqrt{m_1}, \sqrt{m_2})$ with $m_1, m_2 \in \mathbb{Z}$. Let D_i be the discriminant of the field $\mathbb{Q}(\sqrt{m_i})$, $i = 1, 2$.

If p is a rational prime such that $p|D_i$ then

$$p \text{ ramifies in } k_2 \Rightarrow p \text{ ramifies in } F, \text{ as } k_2/F \text{ is unramified} \Rightarrow p|D$$

Now if p divides both D_1 and D_2, then $p|D$ and so p is totally ramified in k_2, which contradicts the fact that k_2/F is unramified. Hence $(D_1,D_2) = 1$. Also if $p|D$, p must divide D_i for $i = 1$ or 2. Hence $D = D_1D_2$. (Note that if both D and D_1 are even, then the power of 2 in D must be the same as the power of 2 in D_1, since otherwise D_2 would also be even contradicting the fact that $(D_1,D_2) = 1$.)

Finally, if $p|D$ and $\mathfrak{p}$ is a prime of F containing p, then the class of $\mathfrak{p}$ is in $C^G \subseteq H$, so by class field theory $\mathfrak{p}$ splits in k_2 and hence p splits in $\mathbb{Q}(\sqrt{D_2})$, so that $(D_2/p) = +1$ where (D_2/p) is the Kronecker symbol.

Therefore $D = D_1D_2$ is a factorization of the second kind.

Conversely, if $D = D_1D_2$ is a factorization of the second kind, let $k_2 = \mathbb{Q}(\sqrt{D_1}, \sqrt{D_2})$.

Then $k_2 \supseteq F$, and since $(D_1,D_2) = 1$ it follows that k_2/F is unramified. If $p|D$, then $p|D_1$, say. Since $D = D_1D_2$ is a factorization of the second kind, then $(D_2/p) = 1$ so p splits in $\mathbb{Q}(\sqrt{D_2})$ and hence $\mathfrak{p}$ splits in k_2, where $\mathfrak{p}$ is a prime of F containing p. Since C^G is generated by the classes of the ramified primes by the lemma, it follows from class field theory that $C^G \subseteq H$, where k_2 is class field to the subgroup $H \subseteq C$, with $[C:H] = 2$.

Hence to the factorization $D = D_1D_2$ we associate the subgroup H, $C^G \subseteq H \subseteq C$, $[C:H] = 2$.

Thus we have a one-one correspondence. Factorizations of second kind $D = D_1D_2 \leftrightarrow k_2 = Q(\sqrt{D_1}, \sqrt{D_2}) \leftrightarrow C^G \subseteq H \subseteq C$, $[C:H] = 2$.

Note that to any factorization $D = D_1D_2$ with $(D_1,D_2) = 1$ we obtain $H \subseteq C$, $[C:H] = 2$.

Hence we have proved

THEOREM 1. If k is the number of invariants of C divisible by 4, then 2^k = the number of factorizations of D of the second kind.

THEOREM 2. If t is the 2-rank of C, then 2^t = the number of factorizations of D, $D = D_1D_2$. Hence $t + 1$ = the number of distinct primes dividing D.

REFERENCE

1. Rédei, L., and Reichardt, H. Die Anzahl der durch 4 teilbaren Invarianten der Klassengruppe eine beliebigen quadratischen Zahlkörpers, J. Reine Angew, Math. 170 (1934), 69-74.

AUTOMORPHS OF TERNARY QUADRATIC FORMS

Wilhelm Plesken*

Department of Mathematics
California Institute of Technology
Pasadena, California

INTRODUCTION

This chapter is concerned with Fricke and Klein's exposition of automorphism groups of integral indefinite ternary quadratic forms as presented in their book "Vorlesungen über die Theorie der automorphen Funktionen" (Band I, dritter Abschnitt, zweites Kapitel: "Von den reproduzierenden Gruppen ternärer und quaternärer quadratischer Formen"). A short survey of this can be found in W. Magnus' book "Non-Euclidean Tesselations and Their Groups," page 133.

The real automorphism group $Aut_{\mathbb{R}}(f)$ of a real quadratic form f (or more precisely of the quadratic space associated with f) is the group of all $g \in GL(n,\mathbb{R})$ with $g^t A_f g = A_f$, where $A_f = (a_{ij})$ is a symmetric $n \times n$ matrix with $f(x_1,\ldots,x_n) = \Sigma^n_{i,j=1} a_{ij} x_i x_j$ for real x_i. In case the a_{ij} are integers, we define the integral automorphism group $Aut_{\mathbb{Z}}(f)$ as the intersection $Aut_{\mathbb{R}}(f) \cap GL(n,\mathbb{Z})$. (Sometimes $Aut_{\mathbb{Z}}(f) \cap SL(n,\mathbb{Z})$ is called the group of units of f.) The elements of $Aut_{\mathbb{R}}(f)$ resp. $Aut_{\mathbb{Z}}(f)$ are called real resp. integral automorphs of f.

In Part I of this chapter we describe the real automorphs of a real ternary quadratic form f which is nondegenerate and indefinite. This is done by using a homomorphism of $SL(2,\mathbb{R})$ into $Aut_{\mathbb{R}}(f)$ and a detailed discussion of the geometric background of this homomorphism. Part II contains a short survey of discontinuous circle groups. Special examples of

*Current affiliation: Lehrstuhl D für Mathematik, Rheinisch-Westfälische Techniche Hochschule, Aachen, Federal Republic of Germany

these are the groups of integral automorphs of ternary quadratic forms, to which Part III is devoted. In particular, a geometric method of finding generators and defining relations for these groups is presented.

I. THE REAL AUTOMORPHS OF A TERNERY QUADRATIC FORM

Let f, g be two real ternary quadratic forms, which are nondegenerate and indefinite. Then there exist an $X \in SL(3,\mathbb{R})$ and a $\lambda \in \mathbb{R} \setminus \{0\}$ such that $\lambda A_f = X^t A_g X$. From this we get at once $\mathrm{Aut}_{\mathbb{R}}(f) = X\ \mathrm{Aut}_{\mathbb{R}}(g) X^{-1}$. Choose in particular g such that $g(z_1,z_2,z_3) = z_1^2 + z_2^2 - z_3^2$. Then $\mathrm{Aut}_{\mathbb{R}}(g)$ is denoted by $O(2,1)$. The following facts about $O(2,1)$ are well known [3]:

(i) $O(2,1)$ is a three-parameter Lie group

(ii) $O(2,1) = SO(2,1) \times \langle -I_3 \rangle$, where $SO(2,1) = O(2,1) \cap SL(3,\mathbb{R})$

(iii) $SO(2,1)$ has two connected components, hence $O(2,1)$ has four connected components.

Clearly, the corresponding statements also hold for $\mathrm{Aut}_{\mathbb{R}}(f)$, where f is as above. We want to establish an isomorphism between $SL(2,\mathbb{R})/\langle -I_2 \rangle$ and the connected 1-component $\mathrm{Aut}^o_{\mathbb{R}}(f)$, where f is given by $f(z_1,z_2,z_3) = z_1 z_3 - z_2^2$. $\mathrm{Aut}^o_{\mathbb{R}}(f)$ is a normal subgroup of $\mathrm{Aut}_{\mathbb{R}}(f)$ of index 4. To this end, we remark that the determinant is a quadratic form on the three-dimensional vector space $S_2(\mathbb{R})$ of symmetric 2×2 matrices,

$$\det\begin{pmatrix} z_1 & z_2 \\ z_2 & z_3 \end{pmatrix} = z_1 z_3 - z_2^2$$

and that for every $g \in SL(2,\mathbb{R})$ and $X \in S_2(\mathbb{R})$ we have $\det(g^t X g) = \det X$.

Let $\mathrm{Aut}^o_{\mathbb{R}}(\det, S_2(\mathbb{R}))$ denote the connected component of the unit element in the automorphism group $\mathrm{Aut}_{\mathbb{R}}(\det, S_2(\mathbb{R}))$ of $S_2(\mathbb{R})$ with the determinant as a quadratic form. Then we shall prove

THEOREM I.1. (i) $\tilde{\Delta}\colon\ SL(2,\mathbb{R})/\langle -I_2 \rangle \to \mathrm{Aut}^o_{\mathbb{R}}(\det, S_2(\mathbb{R}))$ defined by $\tilde{\Delta}(\pm g)X = g^{-t} X g^{-1}$ (with $g^{-t} = (g^t)^{-1}$) for all $g \in SL(2,\mathbb{R})$ is an isomorphism of Lie groups. (ii) Let

$$\phi\colon\ S_2(\mathbb{R}) \to \mathbb{R}_3\colon \begin{pmatrix} z_1 & z_2 \\ z_2 & z_3 \end{pmatrix} \to \begin{pmatrix} z_1 \\ z_2 \\ z_3 \end{pmatrix}$$

and let f denote the ternary quadratic form given by $f(z_1,z_2,z_3) = z_1z_3 - z_2^2$. Let Δ denote the mapping which makes the following diagram commutative

$$\begin{array}{ccc} S_2(\mathbb{R}) & \xrightarrow{\tilde{\Delta}(\pm g)} & S_2(\mathbb{R}) \\ \phi\downarrow & & \downarrow\phi \\ \mathbb{R}_3 & \xrightarrow{\Delta(\pm g)} & \mathbb{R}_3 \end{array} \qquad g \in SL(2,\mathbb{R})$$

Then Δ is an isomorphism of $SL(2,\mathbb{R})/\langle -I_2\rangle$ onto $Aut^o_{\mathbb{R}}(f)$ and

$$\Delta\left(\pm\begin{pmatrix} a & b \\ c & d\end{pmatrix}\right) = \begin{pmatrix} d^2 & -2cd & c^2 \\ -bd & (ad + bd) & -ac \\ b^2 & -2ab & a^2\end{pmatrix} \text{ with } ad - bc = 1$$

Remark. If we had chosen $\tilde{\Delta}(\pm g)X = gXg^t$ in (i), then we would have

$$\Delta\left(\pm\begin{pmatrix} a & b \\ c & d\end{pmatrix}\right) = \begin{pmatrix} a^2 & 2ab & b^2 \\ ac & (ad + bc) & bd \\ c^2 & 2cd & d^2\end{pmatrix}$$

The same formula is also obtained in [1,9,11,12]. The reason for our choice will become clear later. We also sometimes want to consider $\tilde{\Delta}$ and Δ as homomorphisms of $SL(2,\mathbb{R})$. For the proof of the second part of the theorem we shall use the following elementary lemma.

LEMMA I.2. Let $A, X, Y \in M_{n,n}(\mathbb{R})$, and let

$$\bar{X} = \begin{pmatrix} X_1 \\ \vdots \\ X_n\end{pmatrix}, \quad \bar{Y} = \begin{pmatrix} Y_1 \\ \vdots \\ Y_n\end{pmatrix}$$

where X_i (resp. Y_i) are the columns of X (resp. Y). Then

(i) $AX = Y \Leftrightarrow (I_n \otimes A)\bar{X} = \bar{Y}$

(ii) $XA = Y \Leftrightarrow (A^t \otimes I_n)\bar{X} = \bar{Y}$, where $\otimes$ denotes the usual Kronecker product of matrices.

Proof of the theorem: (i) Clearly $\tilde{\Delta}$: $SL(2,\mathbb{R}) \to GL(S_2(\mathbb{R}))$ defined by $\tilde{\Delta}(g)X = g^{-t}Xg^{-1}$ for all $g \in SL(2,\mathbb{R})$, $X \in S_2(\mathbb{R})$ is a homomorphism with kernel $\{\pm I_2\}$. Further, we have $\det(\tilde{\Delta}(g)X) = \det X$, so $\tilde{\Delta}$ can be considered as a monomorphism of $SL(2,\mathbb{R})/\langle -I_2\rangle$ into $Aut_{\mathbb{R}}(\det, S_2(\mathbb{R}))$. As $SL(2,\mathbb{R})$ is connected and $\tilde{\Delta}$ clearly is continuous, the image of $SL(2,\mathbb{R})$ is contained in the connected component $Aut^o_{\mathbb{R}}(\det, S_2(\mathbb{R}))$ of the unit element in $Aut_{\mathbb{R}}(\det, S(\mathbb{R}))$. That it is surjective onto $Aut^o_{\mathbb{R}}(\det, S_2(\mathbb{R}))$ follows from the fact that both $SL(2,\mathbb{R})$ and $Aut_{\mathbb{R}}(\det, S_2(\mathbb{R}))$ are three-parameter Lie groups and that the kernel of $\tilde{\Delta}$ is discrete.

(ii) Let $Z' = \Delta(g)Z = g^{-t}Zg^{-1}$ for $Z \in S_2(\mathbb{R})$ and $g \in SL(2,\mathbb{R})$, then $Z'g = g^{-t}Z$. In the notation of the lemma we get $(g^t \otimes I_2)\bar{Z}' = (I_2 \otimes g^{-t})\bar{Z}$ or $(g^{-t} \otimes g^{-t})\bar{Z} = \bar{Z}'$. With

$$Z = \begin{pmatrix} z_1 & z_2 \\ z_2 & z_3 \end{pmatrix}, \quad g = \begin{pmatrix} a & b \\ c & d \end{pmatrix}, \quad \text{and} \quad g^{-t} = \begin{pmatrix} d & -c \\ -b & a \end{pmatrix}$$

we get

$$\begin{pmatrix} d^2 & -cd & -cd & c^2 \\ -bd & ad & bc & -ac \\ -bd & bc & ad & -ac \\ b^2 & -ab & -ab & a^2 \end{pmatrix} \begin{pmatrix} z_1 \\ z_2 \\ z_2 \\ z_3 \end{pmatrix} = \begin{pmatrix} z_1' \\ z_2' \\ z_2' \\ z_3' \end{pmatrix}$$

or

$$\begin{pmatrix} d^2 & -2cd & c^2 \\ -bd & (ad+bc) & -ac \\ b^2 & -2ab & a^2 \end{pmatrix} \begin{pmatrix} z_1 \\ z_2 \\ z_3 \end{pmatrix} = \begin{pmatrix} z_1' \\ z_2' \\ z_3' \end{pmatrix}$$

proving

$$\Delta\left(\begin{pmatrix} a & b \\ c & d \end{pmatrix}\right) = \begin{pmatrix} d^2 & -2cd & c^2 \\ -bd & (ad+bc) & -ac \\ b^2 & -2ab & a^2 \end{pmatrix}$$

Q.E.D.

Our next aim is a classification of the one-parameter subgroups of $SL(2,\mathbb{R})/\langle -I_2 \rangle$, resp. $Aut_{\mathbb{R}}(f)$. An easy computation yields the following lemma.

LEMMA I.3. The Lie algebra of $Aut_{\mathbb{R}}(f)$ is given by

$$\mathcal{L}(Aut_{\mathbb{R}}(f)) = \{X \in M_{3,3}(\mathbb{R}) \mid X^t A_f + A_f X = 0\}$$

$$= \left\{ \begin{pmatrix} \alpha & 2\beta & 0 \\ \gamma & 0 & \beta \\ 0 & 2\gamma & -\alpha \end{pmatrix} \middle| \alpha, \beta, \gamma \in \mathbb{R} \right\}. \quad \text{Note } A_f = \begin{pmatrix} 0 & 0 & 1/2 \\ 0 & -1 & 0 \\ 1/2 & 0 & 0 \end{pmatrix}.$$

One sees at once that all matrices $\neq 0$ in $L(\mathrm{Aut}(f))$ have rank two. Thus for each $X \in L(\mathrm{Aut}_{\mathbb{R}}(f))$, $X \neq 0$, there exists up to scalar multiples exactly one $Y \in \mathbb{R}_3$ with $Y \neq 0$ and $XY = 0$. The one-parameter subgroup belonging to X is given by $\mathbb{R} \to \mathrm{Aut}_{\mathbb{R}}(f)$: $t \to \exp(tX)$ and has the property $\exp(tX)Y = Y$ for all $t \in \mathbb{R}$, because

$$\exp(tX)Y = \left(\sum_{n=0}^{\infty} \frac{1}{n!} t^n X^n\right) Y = \sum_{n=0}^{\infty} \left(\frac{1}{n!} t^n X^n Y\right) = I_3 Y = Y$$

Together with Theorem I.1 this implies that for every one-parameter subgroup of $SL(2,\mathbb{R})/\langle -I_2\rangle$ there exists up to real multiples exactly one element in $S_2(\mathbb{R})$ (which can be interpreted as a binary quadratic form), being fixed by this one-parameter subgroup acting via the representation $\tilde{\Delta}$.

On the other hand, it is well known that for each $X \in S_2(\mathbb{R})$, $X \neq 0$, there exists a $\lambda \in \mathbb{R}$ and a $g \in SL(2,\mathbb{R})$ such that $\lambda\tilde{\Delta}(g)X = \lambda g^{-tr}Xg^{-1}$ is equal to either of

$$\text{(i)} \begin{pmatrix} 1 & 0 \\ 0 & 1 \end{pmatrix}, \quad \text{(ii)} \begin{pmatrix} 0 & 1 \\ 1 & 0 \end{pmatrix}, \quad \text{or} \quad \text{(iii)} \begin{pmatrix} 1 & 0 \\ 0 & 0 \end{pmatrix}$$

The subgroup of $SL(2,\mathbb{R})/\langle -I_2\rangle$ fixing these elements are (as easy computations show)

$$\text{(i)} \quad E = \left[\begin{pmatrix} \cos\theta & -\sin\theta \\ \sin\theta & \cos\theta \end{pmatrix} \middle| \theta \in \mathbb{R}\right] / \langle -I_2\rangle$$

$$\text{(ii)} \quad H = \left[\pm\begin{pmatrix} e^t & 0 \\ 0 & e^{-t} \end{pmatrix} \middle| t \in \mathbb{R}\right] / \langle -I_2\rangle$$

$$\text{(iii)} \quad P = \left[\pm\begin{pmatrix} 1 & 0 \\ t & 1 \end{pmatrix} \middle| t \in \mathbb{R}\right] / \langle -I_2\rangle$$

respectively. Thus the subgroup of $SL(2,\mathbb{R})/\langle -I_2\rangle$ fixing an $X \in S_2(\mathbb{R})$, $X \neq 0$, is either conjugate to E, H, or P according to whether det X is > 0, < 0, or $= 0$; i.e., whether the real quadratic form represented by X is definite, indefinite and nondegenerate, or degenerate. The subgroups of $SL(2,\mathbb{R})/\langle -I_2\rangle$ conjugate to E, H, or P, respectively, and are called elliptic, hyperbolic, or parabolic, respectively. (Only the elliptic subgroups are compact.) We have proved:

THEOREM I.4. Every subgroup $G \neq \langle I_3\rangle$ of $\mathrm{Aut}_{\mathbb{R}}(f)$ depending continuously on one parameter corresponds to a one-dimensional subspace V of forms in

$S_2(\mathbb{R})$, such that $\phi(V)$ is pointwise invariant under G, and conversely every $V \leq S_2(\mathbb{R})$ with $\dim V = 1$ determines such a subgroup $G \neq \langle I_3 \rangle$ fixing $\phi(V)$.

We aim now at a geometric interpretation of these results. To this end, we introduce projective spaces.

DEFINITION I.5. Let K be a field and V a K-vector space. (In our applications, K is either $\mathbb{R}$, or $\mathbb{C}$ and V is finite dimensional.)

(i) The points of the projective space P(V) corresponding to V are the one-dimensional subspaces of V. The projective subspaces of P(V) with projective dimension k are the (k + 1)-dimensional subspaces of V. In particular, the lines of P(V) are the two-dimensional subspaces of V.

(ii) A projective mapping of P(V) or a projectivity is a mapping α: $P(V) \to P(V)$ which is induced by a nonsingular linear mapping β: $V \to V$; i.e., for all $p \in P(V)$ we have $\alpha(p) = \beta(p) = \{\beta(x) | x \in p\}$. The projectivities of P(V) form a group denoted by PGL(V) which is isomorphic to $GL(V)/\{\lambda \mathrm{id} | \lambda \in K^*\}$.

(iii) If f: $V \to K$ is a quadratic form on V, then $N_f = \{x \in V | f(x) = 0\}$ is called a cone and has the property $x \in N_f \Rightarrow \lambda x \in N_f$ for all $\lambda \in K$. $C_f = \{p \in P(V) | f(x) = 0 \text{ for all } x \in p\}$ is called a conic.

As a first application of these concepts let us look at $P(\mathbb{C}_2)$, with the group of projectivities which are induced by $SL(2,\mathbb{R})$. Clearly this group is isomorphic to $SL(2,\mathbb{R})/\langle -I_2 \rangle$. Consider the mapping ι: $P(\mathbb{C}_2) \to \hat{\mathbb{C}}$, where $\hat{\mathbb{C}}$ is the extended complex plane, defined by

$$\iota\left(\left\{\lambda\begin{pmatrix} z_1 \\ z_2 \end{pmatrix} \middle| \lambda \in \mathbb{C}\right\}\right) = \iota\left(\left\langle\begin{pmatrix} z_1 \\ z_2 \end{pmatrix}\right\rangle\right) = \frac{z_1}{z_2}$$

for all z_1, $z_2 \in \mathbb{C}$ with $|z_1| + |z_2| > 0$. It is easy to see that ι is a bijection. The projectivity induced by

$$g = \begin{pmatrix} a & b \\ c & d \end{pmatrix} \in SL(2,\mathbb{R}) \text{ of } P(\mathbb{C}_2)$$

yields a real Möbius transformation of $\hat{\mathbb{C}}$:

$$\begin{array}{llllll} g\colon & P(\mathbb{C}_2) \to P(\mathbb{C}_2)\colon & \left\langle\begin{pmatrix} z_1 \\ z_2 \end{pmatrix}\right\rangle & \to & \left\langle g\begin{pmatrix} z_1 \\ z_1 \end{pmatrix}\right\rangle = \left\langle\begin{pmatrix} az_1 + bz_2 \\ cz_1 + dz_2 \end{pmatrix}\right\rangle \\ & \iota\downarrow \quad \iota\downarrow & \iota\downarrow & & \iota\downarrow \\ g\colon & \hat{\mathbb{C}} \to \hat{\mathbb{C}}\ : & z & \to & \dfrac{az + b}{cz + b} = gz \end{array}$$

Next consider the elements

$$0 \neq X = \begin{pmatrix} x_1 & x_2 \\ x_2 & x_3 \end{pmatrix} \in S_2(\mathbb{R})$$

as quadratic forms of

$$\mathbb{C}_2: \begin{pmatrix} z_1 \\ z_2 \end{pmatrix} \to (z_1, z_2)X\begin{pmatrix} z_1 \\ z_2 \end{pmatrix} = x_1z_1^2 + 2x_2z_1z_2 + x_3z_2^2$$

The corresponding conic C_X consists of one or two points. More precisely, $\iota(C_X)$ consists of

(i) the two complex conjugate points $(x_2 \pm i\sqrt{\det X})/x_1$ if $\det X > 0$

(ii) the two real points $(x_2 \pm \sqrt{-\det X})/x_1$ for $x_1 \neq 0$, resp. ∞, and $-x_3/2x_2$ for $x_1 = 0$ if $\det X < 0$

(iii) the real point x_2/x_1 for $x_1 \neq 0$ or ∞ for $x_1 = 0$ if $\det X = 0$

as one easily sees by solving a quadratic equation. We remark that for each nonzero real number λ the conics of X and λX in $P(\mathbb{C}_2)$ are equal: $C_X = C_{\lambda X}$. This leads us to the consideration of the real projective space $P(S_2(\mathbb{R}))$ of $S_2(\mathbb{R})$. Further, we point out that by the definition of $\tilde{\Delta}$, resp. the definition of the action of $SL(2,\mathbb{R})$ on $S_2(\mathbb{R})$, we get $C_{\tilde{\Delta}(g)X} = gC_X$ for all $g \in SL(2,\mathbb{R})$ and $X \in S_2(\mathbb{R})\setminus\{0\}$. This follows immediately from the equation

$$(z_1z_2)g^t g^{-t}Xg^{-1}g\begin{pmatrix} z_1 \\ z_2 \end{pmatrix} = (z_1z_2)X\begin{pmatrix} z_1 \\ z_2 \end{pmatrix}$$

Note that $\tilde{\Delta}(g)X = g^{-t}Xg^{-1}$ which provides the reason for our definition of $\tilde{\Delta}$.

Now we want to consider $P(S_2(\mathbb{R}))$, the group of projectivities of $P(S_2(\mathbb{R}))$ which are induced by the action of $SL(2,\mathbb{R})$ and the determinant as a quadratic form on $S_2(\mathbb{R})$. For the sake of brevity let C be the conic in $P(S_2(\mathbb{R}))$ belonging to the determinant. This conic divides $P(S_2(\mathbb{R}))$ into three parts: the conic, its interior, and its exterior. We give a formal definition

DEFINITION I.6. Let V be a real vectorspace of dimension three and f a quadratic form on V with signature (1, 2) (+, -, -). Then f divides P(V) (as well as V) into three parts:

The conic $C_f = \{p \in P(V) \mid f(x) = 0 \text{ for all } x \in p\}$

The interior of C_f: $\{p \in P(V) | f(x) > 0$ for all $x \in p, x \neq 0\}$
The exterior of C_f: $\{p \in P(V)\ f(x) < 0$ for all $x \in p, x \neq 0\}$

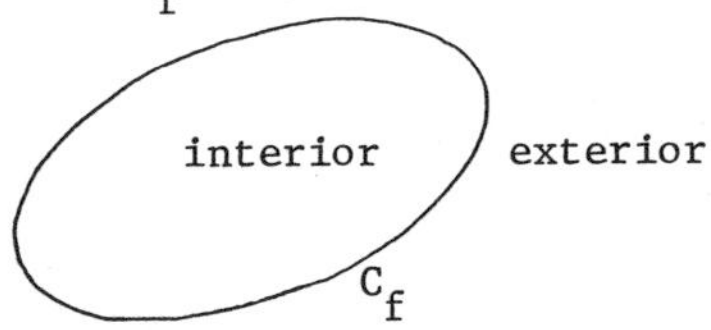

In our situation we get the interior of C as the part of $P(S_2(\mathbb{R}))$ corresponding to the definite forms in $S_2(\mathbb{R})$, C corresponds to the degenerate forms and the exterior of C to the nondegenerate indefinite forms. From an earlier remark (in the proof of Theorem I.4) it follows that $SL(2,\mathbb{R})/\langle -I_2\rangle$ acts on each of these parts transitively. We are particularly interested in the interior of C. As we saw earlier, we could map each definite form X of $S_2(\mathbb{R})$ onto a pair of complex conjugate numbers in $\mathbb{C}$, which were the solution of a quadratic equation of the conic C_X belonging to X. This mapping was constant on the one-dimensional subspaces of $S_2(\mathbb{R})$, so it induces a mapping of the interior of C into the upper complex half plane $\mathbb{C}^u$ in the following way:

α: interior of C $\to$ upper complex half plane $\mathbb{C}^u$ with

$$\alpha\left(\left\langle X = \begin{pmatrix} x_1 & x_2 \\ x_2 & x_3 \end{pmatrix}\right\rangle\right) = \frac{x_2 + i\sqrt{\det X}}{x_1} \text{ for } \det X > 0 \text{ and } x_1 > 0$$

This mapping can easily be extended to carry $C \cup \mathrm{Int}(C)$ into the closed upper half plane in an obvious manner.

Clearly α is surjective, because every complex number occurs as a solution of a real quadratic equation, and α is 1-1, because every pair of complex conjugate numbers a, $\bar{a}$ determines up to a real scalar multiple the quadratic equation, which is satisfied by a and $\bar{a}$. We next have to investigate in which manner the group action of $SL(2,\mathbb{R})/\langle -I_2\rangle$ is carried over to the upper complex half plane by α. But we derive this information immediately from the earlier remark that $C_{\tilde{\Delta}(g)X} = gC_X$ for all $X \in S_2(\mathbb{R})$ and $g \in SL(2,\mathbb{R})$. This implies that the projectivity induced by $g \in SL(2,\mathbb{R})$ on $P(S_2(\mathbb{R}))$, resp. int(C), corresponds to the Möbius transformation of the upper complex half plane $\mathbb{C}^u$, i.e.:

$$g = \begin{pmatrix} a & b \\ c & d \end{pmatrix} \in SL(2,\mathbb{R}) \qquad \begin{array}{ccc} \mathrm{int}(C) & \overset{g}{\to} & \mathrm{int}(C) \\ \alpha\downarrow & & \downarrow \\ \mathbb{C}^u & \underset{g}{\to} & \mathbb{C}^u \end{array} \qquad \begin{array}{ccc} \langle X\rangle & \to & \langle g^{-t}Xg^{-1}\rangle \\ \downarrow & & \downarrow \\ z = \alpha(\langle X\rangle) & \to & \dfrac{az+b}{cz+d} \end{array}$$

Finally, we mention that the projective lines passing through the interior of the conic C are mapped by α onto semicircles in the upper half of the complex plane $\mathbb{C}^u$ which are orthogonal to the real axis (compare [4,9]). So we have the following theorem.

THEOREM I.7. α: $\operatorname{int}(C) \to \mathbb{C}^u$ defined by

$$\alpha\left(\left\langle X = \begin{pmatrix} x_1 & x_2 \\ x_2 & x_3 \end{pmatrix}\right\rangle\right) = \frac{x_2 + i\sqrt{\det X}}{x_1} \quad (x_1 > 0,\ \det X > 0)$$

is a bijection with the following properties:

(i) The segments of the lines of $P(S_2(\mathbb{R}))$ within $\operatorname{int}(C)$ are mapped onto the semicircles orthogonal to the real axis.

(ii) The Möbius transformation induced by $g \in SL(2,\mathbb{R})$ corresponds via α to the transformation of $\langle X\rangle \to \langle g^{-t}Xg^{-1}\rangle$ of $\operatorname{int}(C)$ induced by g.

(iii) Every $g \in SL(2,\mathbb{R})$ induces a projectivity of $P(S_2(\mathbb{R}))$ having exactly one fixed point (for $g \neq \pm I_2$) lying inside C, on C, or outside C according to whether g lies in a elliptic, parabolic, or hyperbolic one-parameter subgroup of $SL(2,\mathbb{R})$.

At this point, we want to make a supplementary remark. Up to now we only dealt with the connected component $\operatorname{Aut}^o_{\mathbb{R}}(f)$ of the unit element in $\operatorname{Aut}_{\mathbb{R}}(f)$, which is of index 4 in $\operatorname{Aut}_{\mathbb{R}}(f)$. More precisely, the factor group is isomorphic to the Klein four group.

Of course, we are interested in the whole $\operatorname{Aut}_{\mathbb{R}}(f)$. We get a subgroup of index 2 of $\operatorname{Aut}_{\mathbb{R}}(f)$ as a homomorphic image of the group

$$SL^{\pm}(2,\mathbb{R}) = \{X \in M_{2\times 2}(\mathbb{R}) \mid \det X = \pm 1\}$$

by extending the homomorphism Δ: $SL(2,\mathbb{R}) \to \operatorname{Aut}_{\mathbb{R}}(f)$ in the following obvious way:

$$\tilde{\Delta}(g)X = g^{-t}Xg^{-1} \text{ for all } g \in SL^{\pm}(2,\mathbb{R}) \text{ and } X \in S_2(\mathbb{R})$$

and

$$\Delta\left(\pm\begin{pmatrix} a & b \\ c & d \end{pmatrix}\right) = \begin{pmatrix} d^2 & -2cd & c^2 \\ -bd & (ad + bc) & -ac \\ b^2 & -2ab & a^2 \end{pmatrix}$$

with $ad - bc = \pm 1$. One sees easily that

$$\det\left(\Delta\begin{pmatrix} a & b \\ c & d \end{pmatrix}\right) = \det\begin{pmatrix} a & b \\ c & d \end{pmatrix}^3 = -1 \text{ if } \det\begin{pmatrix} a & b \\ c & d \end{pmatrix} = -1$$

and that $-I_3 \notin \Delta(SL^{\pm}(2,\mathbb{R}))$. This implies $Aut_{\mathbb{R}}(f) = \Delta(SL^{\pm}(2,\mathbb{R})) \times \langle -I_3 \rangle$. If one defines a new homomorphism Δ': $SL^{\pm}(2,\mathbb{R}) \to Aut_{\mathbb{R}}(f)$ by $\Delta'(g) = \det g\Delta(g)$ for $g \in SL^{\pm}(2,\mathbb{R})$, one gets $\Delta'(SL^{\pm}(2,\mathbb{R})) = Aut_{\mathbb{R}}(f) \cap SL(3,\mathbb{R})$ and $\Delta'(SL^{\pm}(2,\mathbb{R})) \times \langle -I_3 \rangle = Aut_{\mathbb{R}}(f)$.

In order to get the Möbius transformations belonging to the elements in $SL^{\pm}(2,\mathbb{R})$ with determinant -1 that transform the upper complex half-plane $\mathbb{C}^u$ into itself, we have to define

$$\begin{pmatrix} a & b \\ c & d \end{pmatrix}: \mathbb{C}^u \to \mathbb{C}^u: \quad z \to \frac{a\bar{z} + b}{c\bar{z} + d}, \quad \text{where } \det\begin{pmatrix} a & b \\ c & d \end{pmatrix} = -1$$

We mention that $SL^{\pm}(2,\mathbb{R})$ induces the *whole* group of projectivities leaving the conic C_f belonging to f invariant (as a whole).

The interior of C_f together with the group of self mappings induced by $SL^{\pm}(2,\mathbb{R})$ is called the hyperbolic space K_2 [9], the upper complex half plane together with the semicircles orthogonal to the real axis as lines and the Möbius transformations induced by $SL^{\pm}(2,\mathbb{R})$ (see definition above) can be considered as a model for K_2.

At the end of this chapter we want to give a geometric interpretation of the transformation of $\mathbb{R}_3$ carrying an arbitrary indefinite nondegenerate form g into a multiple of f with $f(z_1,z_2,z_3) = z_1z_3 - z_2^2$. We may assume that g is of the same signature than f. Consider the conic C_g in $P(\mathbb{R}_3) = P_2$ belonging to g and let $p = \langle y \rangle$ be a point outside of C_g in P_2. We define the polar line of p and a tangent of C_g.

DEFINITION I.8. Let V, P(V), g, C_g, $p = \langle y \rangle$ as above, and let Φ_g be the scalar product of V corresponding to g, i.e., $\Phi_g(x_1,x_2) = \frac{1}{2}(g(x_1 + x_2) - g(x_1) - g(x_2))$ for $x_1, x_2 \in V$.

(i) The polar line of p with respect to C_g is the two-dimensional subspace

$$\{x \mid x \in V, \ \Phi_g(x,y) = 0\} \text{ of } V$$

(ii) A tangent line to C_g is a line in P(V) having exactly one point in common with C_g.

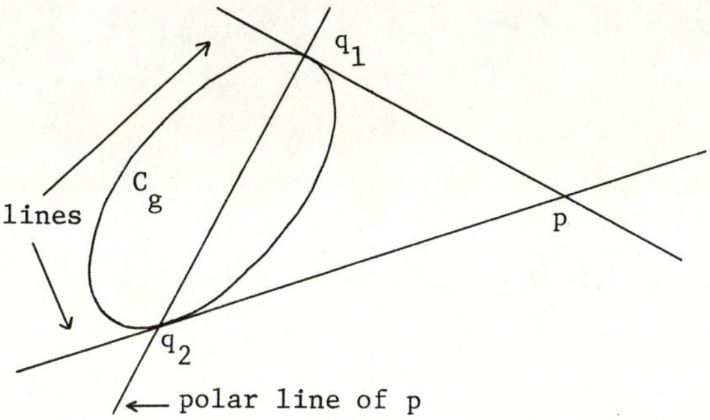

The polar line of p intersects C_g in exactly two points, say q_1 and q_2. The lines passing through p and q_1, resp. p and q_2, are tangent lines to C_g (see [4]. Clearly two different points determine exactly one line passing through them.) On the other side every line intersecting C_g in two points is a polar line of a unique point outside C_g. Assume now that $p_1 = \langle y\rangle$ is an arbitrary point outside C_g and that $q_1 = \langle x\rangle$ and $q_2 = \langle z\rangle$ are the intersection points of the polar line of p with C_g. Then

$$\Phi_g(x,x) = \Phi_g(z,z) = \Phi_g(z,y) = \Phi_g(x,y) = 0$$

By multiplying x and y by appropriate real numbers, we find that

$$\Phi_g(x,z) = \frac{1}{2} \text{ and } \Phi_g(y,y) = -1, \text{ or the matrix } A_g$$

of g with respect to the basis [x,y,z] is equal to

$$A_f = \begin{pmatrix} 0 & 0 & 1/2 \\ 0 & -1 & 0 \\ 1/2 & 0 & 0 \end{pmatrix}$$

We mention that this can be done more elegantly by using the dualism between points and lines in two-dim. projective geometry (see [4]). The following theorem will be of interest later on.

THEOREM I.9 [1]. Let g be an integral ternary quadratic form, which is indefinite and nondegenerate. g represents zero nontrivially, i.e., there exists an $x \in \mathbb{Z}_3$, $x \neq 0$ with $g(x) = 0$, if and only if g is rationally equivalent to a multiple of f with $f(z_1,z_2,z_3) = z_1z_3 - z_2^2$, i.e., there exist a $\lambda \in \mathbb{R}$, $\lambda \neq 0$, and an $h \in GL(3,\mathbb{Q})$ such that $\lambda A_f = h^t A_g h$.

Proof. Let g represent 0 nontrivially. Then there exist an $x \in \mathbb{Z}_3$ such that $g(x) = 0$. Call $\langle x\rangle = \mathbb{R}x = q_1 \in C_g \subseteq P(\mathbb{R}_3)$. Clearly there exists a rational line passing through q_1, intersecting C_g in a second point, where rational means that the two-dim subspace of $\mathbb{R}_3$ corresponding to this line can be defined by a linear equation with rational coefficients. This line intersects C_g in a second point, say $g_2 = \langle z \rangle$, where z can be chosen to be in $\mathbb{Z}_3$. Let $p = \langle y \rangle$ be the point whose polar line is passing through q_1 and q_2. y can be chosen to be rational or integral because it is the solution of a rational system of linear equations. Taking now appropriate integral multiples of x, y, z, one gets A_g corresponding to the basis [x,y,z] as a multiple of A_f. On the other hand, if g does not represent zero nontrivially, then g cannot be rationally equivalent to λf, because λf represents zero nontrivially. □

II. DISCONTINUOUS CIRCLE GROUPS

The study of the integral automorph of a ternary quadratic form (indefinite and nondegenerate) will lead us to discrete subgroups of $SL(2,\mathbb{R})$ acting on the upper complex half-plane $\mathbb{C}^u$ as Möbius transformations. In this chapter we give a short survey on these groups without proving everything. General references are [1], [4], [6], [7], [2] together with [8], and for the first part [10].

Let G^* denote the extension $\langle SL(2,\mathbb{C}), A^* \rangle$ of $SL(2,\mathbb{C})$, where A^* satisfies the relations $A^{*2} = 1$, $A^*gA^{*-1} = \bar{g}$ for all $g \in SL(2,\mathbb{C})$ ($\bar{g}$ = complex conjugate of g) [8].

G^* operates on the extended complex plane by

$$gz = \frac{az + b}{cz + d} \text{ for } g = \begin{pmatrix} a & b \\ c & d \end{pmatrix} \in SL(2,\mathbb{C})$$

and $A^*z = \bar{z}$ (complex conjugation).

The action of G^* on $\hat{\mathbb{C}}$ has the following properties:

(i) Every $g \in G^*$ induces a conformal mapping of $\hat{\mathbb{C}}$.

(ii) Circles of $\hat{\mathbb{C}}$ are mapped onto circles (where straight lines are considered as circles passing through ∞) by the elements of G^*.

(iii) The cross ratio of four (different) points in $\hat{\mathbb{C}}$ is invariant under the action of $SL(2,\mathbb{C})$ and changes into its complex conjugate by applying elements of G^* not lying in $SL(2,\mathbb{C})$, i.e.,

$$\begin{aligned} CR(z_1,z_2,z_3,z_4) &= \frac{z_1 - z_2}{z_2 - z_3}\,\frac{z_3 - z_4}{z_4 - z_1} \\ &= \overline{CR(gz_1,gz_2,gz_3,gz_4)} \qquad \text{for } g \in G^* \setminus SL(2,\mathbb{C}) \\ &= CR(gz_1,gz_2,gz_3,gz_4) \qquad \text{for } g \in SL(2,\mathbb{C}) \end{aligned}$$

(iv) G^* acts transitively on $\hat{\mathbb{C}}$ and the circles of $\hat{\mathbb{C}}$.

DEFINITION II.1. A circle group (Hauptkreisgruppe) is a subgroup of G^* mapping a circle as well as its interior onto itself (if the circle is a straight line, the interior is a half plane).

In order to study circle groups it is enough to restrict the attention to the groups fixing one chosen circle. All other circle groups are conjugate to these, because G^* acts transitively on the circles of $\hat{\mathbb{C}}$. For instance, it would be enough to study the subgroups of

$$SL(2,\mathbb{R}) \cup \{igA^* | g \in GL(2,\mathbb{R}),\ \det g = -1\} \leq G^*$$

which is the biggest subgroup of G^* acting on the real line and the upper half plane $\mathbb{C}^u$.

We now introduce the concept of discreteness of matrix groups and of discontinuity of groups of homeomorphism of topological spaces.

DEFINITION II.2. Let Ω be a topological space and G a group of homeomorphisms of Ω.

(i) $x \in \Omega$ is called a *limit point* of G if there exists a $y \in \Omega$ such that for every neighborhood V of x there exist infinitely many $g \in G$ with $gy \in V$.

(ii) G acts *discontinuously* on $\Omega' \subsetneq \Omega$, where Ω' is an open subset in Ω, if $g\Omega' = \Omega'$ for all $g \in G$ and Ω' does not contain any limit points of G.

(iii) If G is a group of complex matrices, G is called *discrete*, if there is no sequence $\{x_n\}_{n\in\mathbb{N}}$ in G, $x_n \neq$ unit matrix for all $n \in \mathbb{N}$, and $\lim_{n\to\infty} x_n =$ unit matrix.

THEOREM II.3. Let $G \leq SL(2,\mathbb{C})$ be a circle group. G is discrete if and only if G acts discontinuously inside the fixed circle.

Proof. [10], p. 32.

Remark. For convenience, we restrict ourselves to circle groups contained in $SL(2,\mathbb{C})$. The more general circle groups contain these as subgroups of index one and two, because $G^*:SL(2,\mathbb{C}) = 2$.

DEFINITION II.4 [8]. Let G act on Ω' discontinuously. $F \subseteq \Omega'$ is called a fundamental region of G, where F consists of an open set and some of its boundary points, if the following are satisfied:

(i) For every $p \in \Omega'$ there exists a $g \in G$ with $gp \in F$.

(ii) If $p \neq q$ with $p, q \in F$, then $gp \neq q$ for all $g \in G$.

(iii) F is connected.

THEOREM II.5. Let $G \leq SL(2,\mathbb{C})$ be a circle group acting discontinuously inside the circle. Then there exists a fundamental region for G.

Proof. Instead of a proof, we sketch the proof given by Siegel in [10]. In [8], one also finds a proof. We remark that Siegel's definition of fundamental region is slightly different from ours.

(i) $a_1, \ldots, a_4 \in \hat{\mathbb{C}}$ lie on a circle of $\hat{\mathbb{C}}$ if and only if $CR(a_1,a_2,a_3,a_4) \in \mathbb{R}$. $CR(a_1,\ldots,a_4) > 1$ if the a_i are pairwise different and ordered cyclically.

(ii) The cross ratio of four points is invariant under the action of $SL(2,\mathbb{C})$.

(iii) Let K be the circle fixed by G and int(K) the interior of K. Let $z_1 \neq z_2$, z_1, $z_2 \in$ int(K).

Then there is a unique circle $K(z_1,z_2)$ passing through z_1, z_2 and intersecting K orthogonally in two points z_3, z_4.

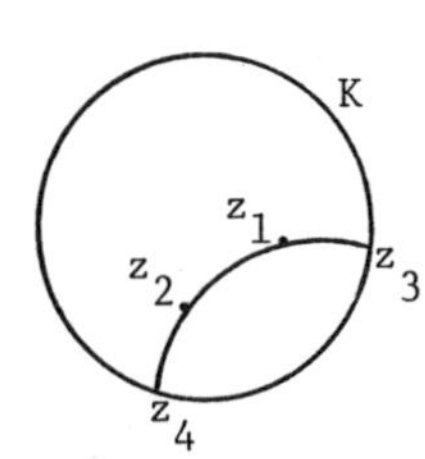

Define

$$\delta(z_1,z_2) = \log(CR(z_1,z_2,z_3,z_4)) \qquad \text{for } z_1 \neq z_2$$

and

$$\delta(z_1,z_2) = 0 \qquad \text{for } z_1 = z_2$$

Then δ has the properties of a metric and is invariant under the action of G. The arcs of the circles intersecting K orthogonally are called lines. With this definition int(K) is called Poincaré's model of non-Euclidean geometry.

(iv) Let z_1, $z_2 \in$ int(K) with $z_1 \neq z_2$. Define $\gamma_{z_1,z_2}(z) = \delta(z_1,z) - \delta(z,z_2)$ for $z \in$ int(K). Then $\gamma_{z_1,z_2}(z) = 0$ defines a line γ_{z_1,z_2} orthogonal to that passing through z_1 and z_2. ℓ_{z_1,z_2} divides int(K) into two parts,

$$B_1(z_1,z_2) = \{z \in \text{int}(K) \mid \gamma_{z_1,z_2}(z) < 0\}$$
$$B_2(z_1,z_2) = \{z \in \text{int}(K) \mid \gamma_{z_1,z_2}(z) \geq 0\}$$

with $z_1 \in B_1(z_1,z_2)$ and $z_2 \in B_2(z_1,z_2)$.

(v) G is countable and there exists an $x_o \in$ int(K) such that the mapping $G/\langle\pm 1\rangle \to$ int(K): $x_o \to gx_o$ is injective. The interior of a fundamental region F of G is given by

$$\overset{o}{F} = \bigcap_{g\in G\setminus\{\pm 1\}} B_1(x_o,gx_o)$$

To get F one has to add certain pieces of certain lines

$\gamma_{x_0,gx_0}(z) = 0$ to $\overset{o}{F}$ Q.E.D.

This completes the proof. We want to consider the case where $\overset{o}{F}$ can already be obtained by taking a finite intersection of the $B_1(x_o,gx_o)$ ($g \in H$, $H \subseteq G$ finite). In this case it is possible to give a finite presentation of G, i.e., to determine a finite set of generators of G with finitely many defining relations.

Aiming at this, we define vertices belonging to the fundamental domain F constructed in the proof of II.5 resp. to the fundamental regions gF with $g \in G$. (Notation as above.)

DEFINITION II.6. $y \in \mathrm{int}(K)$ is called vertex if one of the two conditions is satisfied.

(i) y lies on more than one circle $\gamma_{x_o,gx_o}(z) = 0$

(ii) y lies on one circle $\gamma_{x_o,gx_o}(z) = 0$ and is fixed by an element g of G with $g^2 = \pm 1$ and $g \neq \pm 1$.

An arc $\ell = (z_1,z_2)$ is a line defined by $\gamma_{x_o,gx_o}(z) = 0$ with end points z_1 and z_2 such that z_1 and z_2 are vertices or points on K and such that no other points of ℓ except z_1 or z_2 are vertices.

We restrict ourselves to the case where the boundary of F consists of finitely many arcs.

We give an example of the construction of a fundamental domain before we proceed further.

Example II.7. $SL(2,\mathbb{Z})$ acts on the upper complex half plane $\mathbb{C}^u$ discontinuously.

$$SL(2,\mathbb{Z})/\langle -I_2\rangle \to \mathbb{C}^u: \quad \{\pm g\} \to g(2i)$$

is certainly injective. Let

$$S = \begin{pmatrix} 0 & -1 \\ 1 & 0 \end{pmatrix} \text{ and } T = \begin{pmatrix} 1 & 1 \\ 0 & 1 \end{pmatrix}, \quad \text{then } T^{-1} = \begin{pmatrix} 1 & -1 \\ 0 & 1 \end{pmatrix}$$

Then

$$B_1(2i,S(2i)) = B_1(2i, \tfrac{1}{2}i) = \{z \in \mathbb{C}^u \mid |z| > 1\}$$

$$B_1(2i,T(2i)) = B_1(2i, 2i+1) = \{z \in \mathbb{C}^u \mid \mathrm{Re}(z) < \tfrac{1}{2}\}$$

$$B_1(2i,T^{-1}(2i)) = B_1(2i, 2i-1) = \{z \in \mathbb{C}^u \mid \mathrm{Re}(z) > -\tfrac{1}{2}\}$$

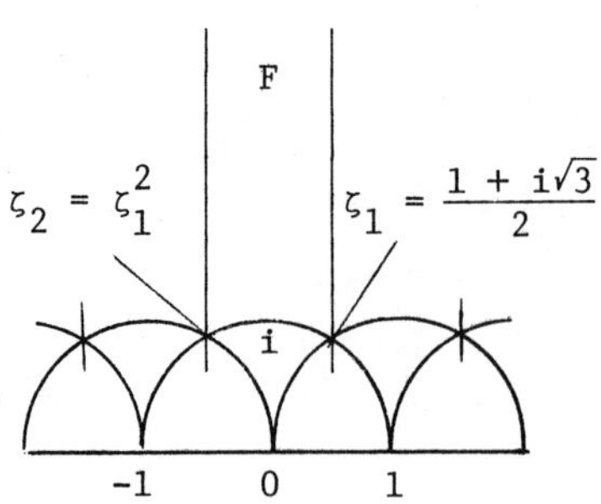

Easy computations show that the intersection $\overset{\circ}{F}$ of these three sets has the property that no two different points of $\overset{\circ}{F}$ lie in the same orbit under $SL(2,\mathbb{Z})$. So $\overset{\circ}{F}$ is already the interior of a fundamental domain F of $SL(2,\mathbb{Z})$. We may choose

$$F = \{z \in \mathbb{C}^u \mid |z| \geq 1,\ -\frac{1}{2} \leq \mathrm{Re}(z) < \frac{1}{2},\ |z| = 1 \Rightarrow \arg z \geq \frac{\pi}{2}\}$$

Vertices of F are ζ_1 and ζ_2, because they are lying on two arcs, and i because Si = i and $S^2 = -1$.

Arcs are the lines (ζ_2,∞), (ζ_1,∞) [equations Re(z) = -1/2, Re(z) = +1/2] and the arcs (ζ_2,i) connecting i with ζ_2 and (i,ζ_1) connecting i with ζ_1.

Note $T(\zeta_2,\infty) = (\zeta_1,\infty)$, $S(\zeta_2,i) = (\zeta_1,i)$.

We return now to the general situation and try to indicate how to get a presentation from a fundamental region F, the boundary of which consists of finitely many arcs. (For details, see [2] and [8, p. 50 ff].) The arcs of F are congruent in pairs under G, which follows from the fact that for each "neighbored" fundamental region F' there is exactly one $g \in G$ mapping F onto F'. Let $\ell_1, \ldots, \ell_k, \ell_{-1}, \ldots, \ell_{-k}$ be the arcs of F such that ℓ_i and ℓ_{-i} are congruent under G and let $g_i \in G$ be the unique element with $g_i\ell_i = \ell_{-i}$. (In the following we consider G to be a subgroup of $SL(2,\mathbb{C})/\langle -I_2\rangle$ rather than of $SL(2,\mathbb{C})$.)

THEOREM II.8. $g_1, \ldots, g_k$ generate G.

Proof. We only give a short sketch. Clearly $\langle g_1,\ldots,g_k\rangle \subseteq G$. So let $h \in G$ and show $h \in \langle g_1,\ldots,g_k\rangle$. Take $x_o \in \mathrm{int}(F)$. Then $G \to \mathrm{int}(K)$: $g \to gx_o$ is injective. Choose a continuous path from x_o to hx_o not containing any vertices. This path is compact and hence can only pass through finitely many images of F under G, say $F = F_1, F_2, \ldots, F_\ell$ in this order.

Certainly there is an $h_1 \in \{g_1,\ldots,g_k,g_1^{-1},\ldots,g_k^{-1}\}$ with $h_1F_1 = F_2$. By the same argument, there will be $h_2 \in \{g_1,\ldots,g_k^{-1}\}$ with $h_2F = h_2h_1^{-1}F_2 = h_1^{-1}F_3$, from which we get $h_1h_2F_1 = F_3$. One can proceed inductively and get, finally, $h_1h_2\cdots h_{\ell-1}F_1 = F_\ell$ with $h_i \in \{g_1,\ldots,g_k^{-1}\}$. Then $h = h_1\cdots h_{\ell-1} \in \langle g_1,\ldots,g_k\rangle$. Q.E.D.

Nearly the same argument yields relations for g_i. Consider a closed simple curve in int(K) such that the interior of this curve contains a vertex. Then the construction of the proof of II.8 would deliver a

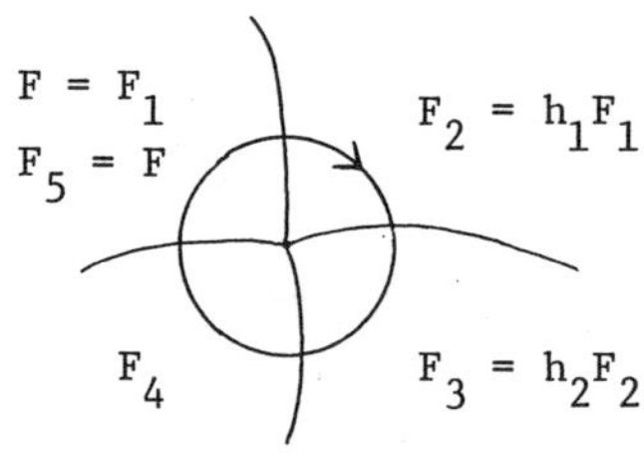

sequence of $h_i \in \{g_1,\ldots,g_k^{-1}\}$ with $h_1h_2\cdots h_{\ell-1}F = F$, i.e., $h_1\cdots h_{\ell-1} = 1$. (Of course, this yields a relation for $G \le SL(2,\mathbb{C})/\langle -I_2\rangle$ and not $G \le SL(2,\mathbb{C})$ because $-I_2z = z$, $\forall\, z \in \mathbb{C}$.) One can prove that one gets a set of defining relations in this way, if one takes a curve around a representative of each of the noncongruent vertices of F. The curves are supposed to pass through only those fundamental regions (congruent to F) having this vertex on this boundary (see [7, p. 50 ff], also [5,6]).

In Example II.7, ζ_1 and i form a set of representatives of noncongruent vertices of F (ζ_1 is congruent to ζ_2). They give rise to the relations $(ST)^3 = \text{id}$ and $S^2 = \text{id}$ (in $SL(2,\mathbb{Z})/\langle -I_2\rangle$), and S and T generate the group.

III. THE INTEGRAL AUTOMORPHISM GROUP OF A TERNARY QUADRATIC FORM

In a previous talk of this seminar, H. Zassenhaus proved the fundamental theorem by making use of the Gauss-Dedekind-Minkowski reduction theory and [13]: Let f be a nondegenerate integral quadratic form. Then $\text{Aut}_{\mathbb{Z}}(f)$ is a finitely presentable discrete subgroup of $\text{Aut}_{\mathbb{R}}(f)$ with compact factor space. This implies, of course, that $\text{Aut}_{\mathbb{Z}}(f)$ is infinite where f is a ternary nondegenerate indefinite integral form that is infinite because $\text{Aut}_{\mathbb{R}}(f)$ is not compact in this case. In this chapter we shall try to give some results on the arithmetic structure of $\text{Aut}_{\mathbb{Z}}(f)$ and to discuss methods of finding a finite presentation. First we give an example. example.

Example III.1. Consider the form f with $f(z_1,z_2,z_3) = 2(z_1z_3 - z_2^2)$. This is an integral form with matrix

$$A_f = \begin{pmatrix} 0 & 0 & 1 \\ 0 & -2 & 0 \\ 1 & 0 & 0 \end{pmatrix}$$

It follows immediately from Theorem I.1 that $\mathrm{Aut}^o_{\mathbb{Z}}(f) = \mathrm{Aut}_{\mathbb{Z}}(f) \cap \mathrm{Aut}^o_{\mathbb{R}}(f)$, the intersection of $\mathrm{Aut}_{\mathbb{Z}}(f)$ with the 1-component of $\mathrm{Aut}_{\mathbb{R}}(f)$ is isomorphic to $SL(2,\mathbb{Z})/\langle \pm I_2 \rangle$ and from II.7 we get

$$V_1 = \Delta\left(\pm\begin{pmatrix} 1 & 1 \\ 0 & 1 \end{pmatrix}\right) = \begin{pmatrix} 1 & 0 & 0 \\ -1 & 1 & 0 \\ 1 & -2 & 1 \end{pmatrix}$$

and

$$V_2 = \Delta\left(\pm\begin{pmatrix} 0 & -1 \\ 1 & 0 \end{pmatrix}\right) = \begin{pmatrix} 0 & 0 & 1 \\ 0 & -1 & 0 \\ 1 & 0 & 0 \end{pmatrix}$$

as generators of $\mathrm{Aut}^o_{\mathbb{Z}}(f)$ with defining relations

$$(V_1V_2)^3 = V_2^2 = I_3$$

According to the remark following I.7, $\mathrm{Aut}_{\mathbb{Z}}(f) \cap SL(3,\mathbb{Z})$ is isomorphic to $SL^{\pm}(2,\mathbb{Z})$. We get a fundamental domain for $SL^{\pm}(2,\mathbb{Z})$ acting on the upper complex half plane quite easily from the fundamental region F of $SL(2,\mathbb{Z})$ because $SL(2,\mathbb{Z})$ is of index two in $SL^{\pm}(2,\mathbb{Z})$. We simply have to divide F given in II.7 into two parts

$$F_1 = F \cap \{z \in C \mid \mathrm{Re}\, z \leq 0\} \qquad \text{and} \qquad F_2 = F \cap \{z \in \mathbb{C} \mid \mathrm{Re}\, z \geq 0\}$$

each of which is a fundamental region for $SL^{\pm}(2,\mathbb{Z})$. (Note $\begin{pmatrix} -1 & 0 \\ 0 & 1 \end{pmatrix}$: $z \to -\bar{z}$ interchanges F_1 and F_2.) (Compare [4].) We get

$$\tilde{V}_1 = \Delta'\left(\pm\begin{pmatrix} -1 & 0 \\ 0 & 1 \end{pmatrix}\right) = -\begin{pmatrix} 1 & 0 & 0 \\ 0 & -1 & 0 \\ 0 & 0 & 1 \end{pmatrix} = \begin{pmatrix} -1 & 0 & 0 \\ 0 & 1 & 0 \\ 0 & 0 & -1 \end{pmatrix}$$

$$\tilde{V}_2 = \Delta'\left(\pm\begin{pmatrix} -1 & 0 \\ 0 & 1 \end{pmatrix}\begin{pmatrix} 0 & -1 \\ 1 & 0 \end{pmatrix}\right) = \Delta'\left(\pm\begin{pmatrix} 0 & 1 \\ 1 & 0 \end{pmatrix}\right) = -\begin{pmatrix} 0 & 0 & 1 \\ 0 & 1 & 0 \\ 1 & 0 & 0 \end{pmatrix} = \begin{pmatrix} 0 & 0 & -1 \\ 0 & -1 & 0 \\ -1 & 0 & 0 \end{pmatrix}$$

$$\tilde{V}_3 = \Delta'\left(\pm\begin{pmatrix} 1 & 1 \\ 0 & 1 \end{pmatrix}\begin{pmatrix} -1 & 0 \\ 0 & 1 \end{pmatrix}\right) = \Delta'\left(\pm\begin{pmatrix} -1 & 1 \\ 0 & 1 \end{pmatrix}\right) = -\begin{pmatrix} 1 & 0 & 0 \\ -1 & -1 & 0 \\ 1 & 2 & 1 \end{pmatrix} = \begin{pmatrix} -1 & 0 & 0 \\ 1 & 1 & 0 \\ -1 & -2 & -1 \end{pmatrix}$$

satisfying the relations

$$(\tilde{V}_1)^2 = (\tilde{V}_2)^2 = (\tilde{V}_3)^2 = (\tilde{V}_2\tilde{V}_1)^2 = (\tilde{V}_3\tilde{V}_2)^3 = I_3$$

as generators for $SL(3,\mathbb{Z}) \cap Aut_{\mathbb{Z}}(f)$. (For the definition of Δ' compare the remark after I.7.) Together with $\tilde{V}_4 = -I_3$ we get a set of generators for $Aut_{\mathbb{Z}}(f)$.

As for $Aut^o_{\mathbb{Z}}(f) \cong SL(2,\mathbb{Z})/\langle -I_2\rangle$ we shall see that there is a one-to-one correspondence between those one-dimensional subspaces of $S_2(\mathbb{Q})$ (the space of rational symmetrix 2×2 matrices) whose elements $\neq 0$ have negative determinant and the maximal cyclic subgroups of $SL(2,\mathbb{Z})/\langle -I_2\rangle$ contained in a one-parameter hyperbolic subgroup of $SL(2,\mathbb{R})/\langle -I_2\rangle$. This is a discrete version of a part of Theorem I.4, which can easily be proved by the theorem quoted at the beginning of this chapter. (The elements of $S_2(\mathbb{Q})$ can be considered as multiples of integral binary forms.)

We now give some general results.

THEOREM III.2. Let f be a ternary integral form nondegenerate and indefinite. Then $Aut^o_{\mathbb{Z}}(f)$, the intersection of $Aut_{\mathbb{Z}}(f)$ with the one-component of $Aut_{\mathbb{R}}(f)$, acts discontinuously on the upper complex half plane $\mathbb{C}^u$ and has a fundamental region there.

Proof. Since $Aut^o_{\mathbb{Z}}(f) \subseteq SL(3,\mathbb{Z})$ is a discrete group, so is $X^{-1}Aut_{\mathbb{Z}}(f)X$, where $X \in GL(3,\mathbb{R})$ such that

$$X^tA_fX = \begin{pmatrix} 0 & 0 & 1/2 \\ 0 & -1 & 0 \\ 1/2 & 0 & 0 \end{pmatrix}$$

Hence by Theorem I.1 the inverse image of $X^{-1}Aut_{\mathbb{Z}}(f)X$ under Δ is a discrete subgroup of $SL(2,\mathbb{R})$. By II.3 this group acts discontinuously on the upper half plane and by II.4 it has a fundamental region there. Q.E.D.

We now want to consider the problem of how much the automorphism groups of rationally equivalent forms differ from one another.

DEFINITION III.3. Let G be a group with subgroups H_1 and H_2. H_1 and H_2 are called *commensurable*, if H_1 can be conjugated into a subgroup $\bar{H}_1$ of G (by an element of G) such that the indices $[\bar{H}_1 : \bar{H}_1 \cap H_2]$ and $[H_2 : \bar{H}_1 \cap H_2]$ are finite.

THEOREM III.4. (For binary forms compare [5], for ternary form [1].) Let f and g be rationally equivalent forms of degree n. Then $\mathrm{Aut}_{\mathbb{Z}}(f)$ and $\mathrm{Aut}_{\mathbb{Z}}(g)$ are commensurable (as subgroups of $GL(n,\mathbb{Q})$).

Proof. Let $X \in GL(n,\mathbb{Q})$ such that $A_f = X^{tr}A_gX$. Then $X\,\mathrm{Aut}_{\mathbb{Q}}(f)X^{-1} = \mathrm{Aut}_{\mathbb{Q}}(g)$. Without loss of generality, we may assume that X is integral, i.e., $X \equiv 0(\mathbb{Z})$. For any lattice L in $\mathbb{Q}^n$, let

$$\mathrm{Aut}(L) = \{h \in GL(n,\mathbb{Q})\,|\,hL = L\}$$

so in particular $\mathrm{Aut}(\mathbb{Z}^n) = GL(n,\mathbb{Z})$. From this we get

$$\mathrm{Aut}_{\mathbb{Z}}(f) = \mathrm{Aut}_{\mathbb{Q}}(f) \cap \mathrm{Aut}(\mathbb{Z}^n)$$

Hence

$$X\,\mathrm{Aut}_{\mathbb{Z}}(f)X^{-1} = \mathrm{Aut}_{\mathbb{Q}}(g) \cap \mathrm{Aut}(X\mathbb{Z}^n)$$

and also

$$\mathrm{Aut}_{\mathbb{Z}}(g) = \mathrm{Aut}_{\mathbb{Q}}(g) \cap \mathrm{Aut}(\mathbb{Z}^n)$$

We have to show that $[\mathrm{Aut}_{\mathbb{Z}}(g)\colon\ \mathrm{Aut}_{\mathbb{Z}}(g) \cap X\,\mathrm{Aut}_{\mathbb{Z}}(f)X^{-1}]$ and $[X\,\mathrm{Aut}_{\mathbb{Z}}(g)X^{-1}\colon \mathrm{Aut}_{\mathbb{Z}}(g) \cap X\,\mathrm{Aut}_{\mathbb{Z}}(f)X^{-1}]$ are finite. For this it is enough to show that

$$[\mathrm{Aut}(\mathbb{Z}^n)\colon\ \mathrm{Aut}(\mathbb{Z}^n) \cap \mathrm{Aut}(X\mathbb{Z}^n)]$$

and

$$[\mathrm{Aut}(X\mathbb{Z}^n)\colon\ \mathrm{Aut}(\mathbb{Z}^n) \cap \mathrm{Aut}(X\mathbb{Z}^n)]$$

are finite.

We show that the first index is finite. Let d = det X. Then $d \neq 0$, $\mathrm{Aut}(\mathbb{Z}^n)$ leaves $d\mathbb{Z}^n$ invariant, and $d\mathbb{Z}^n \subseteq X\mathbb{Z}^n$. We get a canonical homomorphism of $\mathrm{Aut}(\mathbb{Z}^n)$ to $\mathrm{Aut}(\mathbb{Z}^n/d\mathbb{Z}^n)$, because every automorphism of $\mathbb{Z}^n$ induces an automorphism of the finite abelian group $\mathbb{Z}^n/d\mathbb{Z}^n$. An easy verification shows that $\mathrm{Aut}(\mathbb{Z}^n) \cap \mathrm{Aut}(X\mathbb{Z}^n)$ is the inverse image of the subgroup of $\mathrm{Aut}(\mathbb{Z}^n/d\mathbb{Z}^n)$ leaving $X\mathbb{Z}^n/d\mathbb{Z}^n$ (as a whole) invariant. Because $\mathrm{Aut}(\mathbb{Z}^n/d\mathbb{Z}^n)$ is finite, the homomorphism theorem yields that $[\mathrm{Aut}(\mathbb{Z}^n)\colon \mathrm{Aut}(\mathbb{Z}^n) \cap \mathrm{Aut}(X\mathbb{Z}^n)]$ is finite.

The proof of the finiteness of $[\mathrm{Aut}(X\mathbb{Z}^n)\colon\ \mathrm{Aut}(X\mathbb{Z}^n) \cap \mathrm{Aut}(\mathbb{Z}^n)]$ is the same, if one uses $\mathrm{Aut}(\mathbb{Z}^n) = \mathrm{Aut}(d\mathbb{Z}^n)$. Q.E.D.

This theorem is important for constructions of fundamental regions for groups of automorphs of rationally equivalent ternary forms. Let f and g be two rationally equivalent ternary forms, transform $\mathrm{Aut}^{o}_{\mathbb{Z}}(f)$ and $\mathrm{Aut}^{o}_{\mathbb{Z}}(g)$ into the $\mathbb{R}$-automorphism group of $z_1z_3 - z_2^2$ and consider the inverse image of the transformed groups under Δ (compare I.1) in $SL(2,\mathbb{R})$ and denote these groups by G_f and $G_g \le SL(2,\mathbb{R})$. Both of them have fundamental domains. Because of the commensurability of $\mathrm{Aut}_{\mathbb{Z}}(f)$ and $\mathrm{Aut}_{\mathbb{Z}}(g)$ resp. of $\mathrm{Aut}^{o}_{\mathbb{Z}}(f)$ and $\mathrm{Aut}^{o}_{\mathbb{Z}}(g)$, there is a subgroup H of $SL(2,\mathbb{R})$ which is contained in G_f and G_g with a finite index. From a fundamental region F of G_f we get one for H by taking $\tilde{F} = \cup_{i=1}^{k} h_i F$, where the h_i form an appropriate set of coset representatives of H in G_f. Then $\tilde{F}$ is divided into $[G_g : H]$ parts under the action of G_g, each of which is a fundamental region for G_g. We shall give an example of this procedure later on. If, for instance, G_f and $SL(2,\mathbb{Z})$ are commensurable, we say $\mathrm{Aut}_{\mathbb{Z}}(f)$ is commensurable with $SL(2,\mathbb{Z})$.

THEOREM III.5. Let f be a ternary nondegenerate indefinite integral form. If f represents zero nontrivially, then $\mathrm{Aut}_{\mathbb{Z}}(f)$ is commensurable with the modular group $SL(2,\mathbb{Z})$.

Proof. As f represents zero nontrivially, it is rationally equivalent to $z_1z_3 - z_2^2$ by Theorem I.9. By Example III.1,

$$\mathrm{Aut}^{o}_{\mathbb{Z}}(z_1z_3 - z_2^2) = SL(2,\mathbb{Z})/\langle -I_2 \rangle$$

so $\mathrm{Aut}_{\mathbb{Z}}(f)$ is commensurable with $SL(2,\mathbb{Z})$. Q.E.D.

Before we consider examples, we give a result on the elements of $\mathrm{Aut}_{\mathbb{Z}}(f)$, which is related to I.4 and can be used as described in [1], p. 511, to construct cyclic subgroups of $\mathrm{Aut}_{\mathbb{Z}}(f)$.

THEOREM III.6 [1,11,12]. Let $g \in \mathrm{Aut}_{\mathbb{Z}}(f)$. Then 1 is eigenvalue of g if det g = 1 and -1 if det g = -1.

Proof. Let $g \in \mathrm{Aut}^{o}_{\mathbb{Z}}(f)$. Then according to I.1 there is a $\tilde{g} \in SL(2,\mathbb{R})$ and $X \in GL(3,\mathbb{R})$ with $X^{-1}\Delta(\tilde{g})X = g$. $\Delta(\tilde{g})$ was obtained from the Kronecker product of $\tilde{S}$ by reduction. If α and β are the eigenvalues of $\tilde{g}$, then α^2, $\alpha\beta$, $\beta\alpha$, β^2 are the eigenvalues of $\tilde{g} \otimes \tilde{g}$. But det $\tilde{g}$ = 1 and g is real so $\alpha\beta = 1$. Hence 1 is eigenvalue $\Delta(\tilde{g})$, i.e., of g. If det g = 1 and $g \in \mathrm{Aut}_{\mathbb{Z}}(f)$ without being in $\mathrm{Aut}^{o}_{\mathbb{Z}}(f)$, the proof is the same by using the remark following I.7. If det g = -1 then det(-g) = 1 and hence has eigenvalue +1, so g has eigenvalue 1. Q.E.D.

Remark. Clearly, the same holds for $\mathrm{Aut}_{\mathbb{R}}(f)$.

We now consider examples which solve the problem of finding the groups of $\mathbb{Z}$-automorphs of all ternary quadratic forms to a certain degree. From the theory of quadratic forms over fields it is known that every such form is equivalent to a diagonal form. In our situation this means that every integral ternary indefinite and nondegenerate form f or its negative -f is rationally equivalent to the special form denoted by [p,q,r] given by $[p,q,r](z_1,z_2,z_3) = pz_1^2 - qz_2^2 - rz_3^2$. We may further assume that p, q, and r are pairwise relatively prime. For instance, p and q have a common divisor s. Then [p,q,r] and [sp,sq,sr] have the same automorph and by the rational substitution $z_1^1 = (1/s)z_1$, $z_2^1 = (1/s)z_2$, $z_3^1 = z_3$, this is rationally equivalent to [p/s,q/s,sr] where (p/s,q/s) = 1. Similarly p, q, and r can be assumed to be square free. We mention that it is known from number theory (compare Gauss, Disquisitiones arithmeticae, Art. 294 ff) that [p,q,r] represents zero nontrivially iff -pq, -pr, and -qr are quadratic residues (mod, r, q, p, respectively).

We want to determine the automorphism group of [p,q,r]. Let X be a real matrix transforming [p,q,r] in $z_1z_3 - z_2^2$. Then

$$\Delta_X:\quad SL(2,\mathbb{R})/\langle -I_2\rangle \to \mathrm{Aut}^{o}_{\mathbb{R}}([p,q,r]):\quad \{\pm g\} \to X^{-1}\Delta(g)X$$

where Δ, as in Theorem I.1, is an isomorphism. By this one proves by elementary verification the following lemma.

LEMMA III.7. $\Delta_X:\ SL(2,\mathbb{R})/\langle -I_2\rangle \to \mathrm{Aut}^{o}_{\mathbb{R}}([p,q,r])$, given by

$$\Delta_X\left\{\pm\begin{pmatrix} a & b \\ c & d\end{pmatrix}\right\} = \begin{bmatrix} \frac{1}{2}(a^2+b^2+c^2+d^2) & -\sqrt{\frac{q}{p}}(ab+cd) & \frac{1}{2}\sqrt{\frac{r}{q}}(-a^2+b^2-c^2+d^2) \\ -\sqrt{\frac{p}{q}}(ac+db) & (ad+bc) & \sqrt{\frac{r}{q}}(ac-bd) \\ \frac{1}{2}\sqrt{\frac{p}{r}}(-a^2-b^2+c^2+d^2) & \sqrt{\frac{q}{r}}(ad-bc) & \frac{1}{2}(a^2-b^2-c^2+d^2) \end{bmatrix}$$

is an isomorphism of $SL(2,\mathbb{R})/\langle -I_2\rangle$ onto the 1-component $\mathrm{Aut}^{o}_{\mathbb{R}}([p,q,r])$ $\mathrm{Aut}_{\mathbb{R}}([p,q,r])$.

It follows immediately from this lemma that $SL(2,\mathbb{R})/<-I_2>$ is isomorphic to $Aut^{o}_{\mathbb{R}}([1,1,1])$. We also mention that the isomorphism in III.7 can easily be extended to an isomorphism of $SL^{\pm}(2,\mathbb{R})/<-I_2>$ onto $SL(3,\mathbb{R}) \cap Aut_{\mathbb{R}}([p,q,r])$ according to the remark following I.2.

We now look at the forms [1,q,1] (which certainly represent zero nontrivially) and ask for the elements of $SL(2,\mathbb{R})$ whose images under Δ_X are in $Aut_{\mathbb{Z}}([1,q,1])$. We state without proof the following theorem, which is proved in [1] more generally. (Compare also [9].)

THEOREM III.8. Let q be square free and let $q = q_1q_2$ be any decomposition of q into the product of two positive integers q_1, q_2. Then the Möbius transformation $z' = (az + b)/(cz + d)$ defines a unit of [1,q,1] if and only if

(i) q is even and $a = \alpha\sqrt{q_1}$, $b = \beta\sqrt{q_2}$, $c = \gamma\sqrt{q_2}$, $\delta = d\sqrt{q_1}$, with α, β, γ, $\delta \in \mathbb{Z}$, and $q_1\alpha\delta - q_2\beta\gamma = 1$.

(ii) q is odd and either $a = \alpha\sqrt{q_1}$, $b = \beta\sqrt{q_2}$, $c = \gamma\sqrt{q_2}$, $d = \delta\sqrt{q_1}$, $\alpha + \beta + \gamma + \delta = 0 \pmod 2$, $q_1\alpha\delta - q_2\beta\gamma = 1$, or

$$a = \alpha\sqrt{q_1/2}, \quad b = \beta\sqrt{q_2/2}, \quad c = \gamma\sqrt{q_2/2}, \quad d = \delta\sqrt{q_1/2}$$

and $q_1\alpha\delta - q_2\beta\gamma = 2$. In both cases, α, β, γ, $\delta \in \mathbb{Z}$.

Some elementary verification yields the next theorem which allows a general approach to the construction of fundamental regions for $Aut_{\mathbb{Z}}([1,q,1])$. (See [1] and [9].)

THEOREM III.9. The Möbius transformations

$$z' = \frac{az + b\sqrt{g}}{c\sqrt{qz} + d} \qquad (ad - qbc = 1)$$

(where $a + b + c + d \equiv 0 \pmod 2$ if q is odd) define a normal subgroup H of finite index in $\Delta_X^{-1}(Aut^{o}_{\mathbb{Z}}[1,q,1])$. The group H is called the principal subgroup of $\Delta_X^{-1}(Aut^{o}_{\mathbb{Z}}[1,q,1])$. It is conjugate with the subgroup H_q given by

$$H_q = \left\{ \begin{pmatrix} a & b \\ qc & d \end{pmatrix} \middle| ad - qbc = 1 \right\}$$

$$H_q = DHD^{-1} \text{ with } D = \begin{bmatrix} \sqrt[4]{q} & 0 \\ 0 & \frac{1}{\sqrt[4]{q}} \end{bmatrix}$$

H_q is clearly of finite index in SL(2,$\mathbb{Z}$). This allows us to construct a fundamental region for H_q by taking several images of a fundamental region of SL(2,$\mathbb{Z}$). D: $z \to \sqrt{q}z$ maps this fundamental region onto a fundamental region F_H of H. F_H then is divided again in finitely many parts, each of which constitutes a fundamental region for $\Delta_X^{-1}(\mathrm{Aut}_{\mathbb{Z}}^{o}[1,q,1])$. We give a sketch of an example.

Example III.10. Consider [1,2,1]. H_2 is of index 6 in $SL^{\pm}(2,\mathbb{Z})$. We get a fundamental domain for H_2 as indicated in the following figure (compare [4], p. 289).

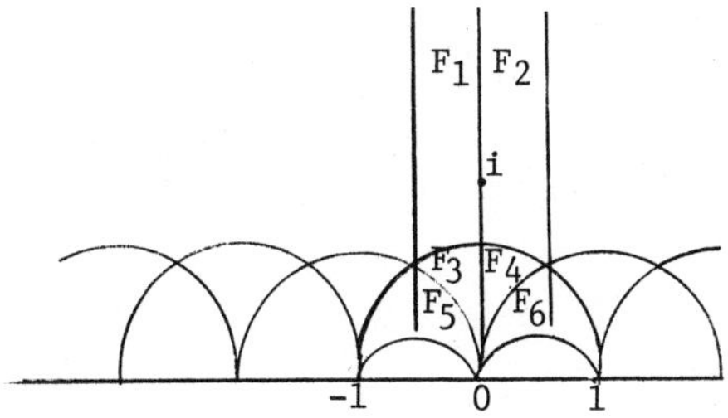

F_i = fundamental region for $SL^{\pm}(2,\mathbb{Z})$ (compare III.1). Six of these fundamental domains join to form one for H_2. Map $F' = \cup_{i=1}^{6} F_i$ onto $F = \sqrt{2}F'$. This is a fundamental region for H. F is now divided into four parts under the action of SL(3,$\mathbb{Z}$) $\cap$ Aut([1,2,1]) according to the next figure.

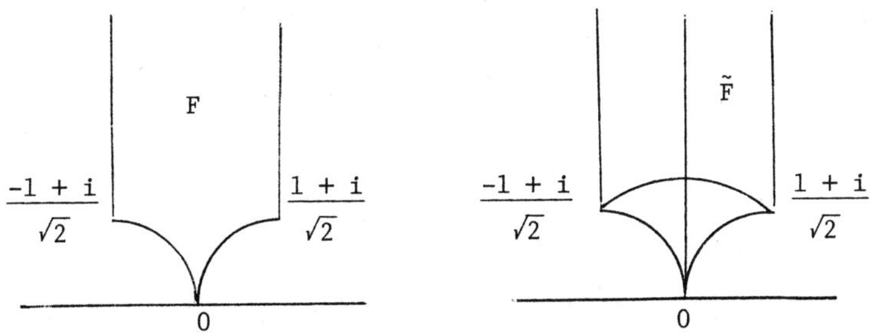

$\tilde{F}$ is a fundamental region for this group. (Remark: All circles are orthogonal to the real line.) For details, compare [5], p. 289, and [1], p. 548. In [1], p. 448 ff, one also finds many examples of forms which do not present zero nontrivially. For a list, compare also [9].

Acknowledgment: I would like to thank D. Shapiro for reading the manuscript and suggesting many improvements.

REFERENCES

1. Fricke, R., and Klein, F. (1897). "Vorlesungen über die Theorie der automorphen Funktionen," Vol. 1, Teubner, Leibzig [Reprinted by Johnson Reprint, New York, 1965]

2. Ford, R. L. (1951). "Automorphic Functions," Chelsea, Bronx, New York

3. Helgason, S. (1962). "Differential Geometry and Symmetric Spaces," Academic Press, New York, London

4. Klein, F. (1928). "Vorlesungen über nicht-euklidische Geometrie," Springer-Verlag, Berlin, Heidelberg, New York

5. Klein, F., and Fricke, R. (1890). "Vorlesungen über die Theorie der elliptischen Modulfunktionen," Vol. 1, Teubner Leipzig [Reprinted by Johnson Reprint, New York, 1965]

6. Lehner, J. (1964). "Discontinuous Groups and Automorphic Functions," Math. Surveys No. 8, Amer. Math. Soc., Providence, Rhode Island

7. Macbeath, A. M. (1965). "Fuchsian Groups," Birmingham (Notes)

8. Magnus, W. (1952). "Discrete Groups," New York University (Notes)

9. Magnus, W. (1974). "Noneuclidean Tesselations and Their Group," Academic Press, New York and London

10. Siegel, C. L. (1971). "Topics in Complex Function Theory," Vol. 2, Wiley (Interscience), New York

11. Taussky, O. (1968). "Automorphs and Generalized Automorphs of Quadratic Forms Treated as Characteristic Value Relations," Linear Algebra and Its Applications 1, 349-356

12. Taussky, O. (1972). "Automorphs of Quadratic Forms as Positive Operators," Inequalities, Vol. III, 341-345

13. Zassenhaus, H. (1971). "On the Units of Orders," Journal of Algebra 20, 368-395

ON A THEOREM OF GAUSZ CONCERNING THE NUMBER OF INTEGRAL SOLUTIONS OF THE EQUATION $x^2 + y^2 + z^2 = m$

Hans Peter Rehm*

California Institute of Technology
Department of Mathematics
Pasadena, California

I. INTRODUCTION

This lecture contains Venkow's proof of the following Theorem 1 of Gausz's, more precisely: a shorter version in modernized language. It is based on Venkow's paper [4], part II.

THEOREM 1. Let m be a square-free integer, $m > 1$, and $m \equiv 1, 2 \pmod 4$, h(m) the ideal class number of the maximal order of the quadratic field $\mathbb{Q}(\sqrt{-m})$, $\psi(m)$ the number of vectors $(x,y,z) \in \mathbf{Z}^3$ such that $x^2 + y^2 + z^2 = m$. Then there holds $\psi(m) = 12 \cdot h(m)$.

The Venkow proof is interesting with respect to the following features:

(i) It uses elementary arithmetic of the (Hurwitz) quaternions.

(ii) It gives an interpretation of the ideal group as a permutation group of an easily computed finite set.

(iii) The proof is constructive, yielding representatives of the ideal classes.

It should be remarked that this lecture covers only a slight, but substantial part of the results given in Venkow's papers [4]. In particular, there is the same result for the case $m \equiv 3 \pmod 8$, if we replace the factor 12 in Gausz's theorem by 24. (The case $m \equiv 7 \pmod 8$ cannot

*Current affiliation: Mathematisches Institutä II, Universität KARLSRUHE, West Germany

be treated by Hurwitz quaternions, since then $\mathbb{Q}(\sqrt{-m})$ cannot be embedded.)

Similar results hold for the group of classes of regular ideals in an order that is not the maximal order of the quadratic number field we are concerned with (Gausz [2], Venkow [4], part III); that is, if we allow m to contain square factors.

As an easy reference for the simple facts from the theory of Hurwitz quaternions which are used here without proof, I recommend Hurwitz [3].

The exposition is as elementary as possible.

II. REVIEW OF HURWITZ QUATERNIONS

a. Some Algebraic Properties

The *Hurwitz quaternion field* $\mathbf{H}$ is the algebra over the field $\mathbb{Q}$ of rational numbers with the vector space basis 1, i_1, i_2, i_3, multiplied according to Hamilton's rule, $i_1 i_2 = -i_2 i_1 = i_3, i_2 i_3 = -i_3 i_2 = i_1, i_3 i_1 - i_1 i_3 = i_2$, $1 \cdot i_r = i_r \cdot 1 = i_r (r = 1,2,3)$. For each quaternion $\alpha = a_0 + a_1 i_1 + a_2 i_2 + a_3 i_3$, we define the *conjugated* quaternion $\bar{\alpha} = a_0 - a_1 i_1 - a_2 i_2 - a_3 i_3$, the *norm* $n(\alpha) = \alpha\bar{\alpha} = a_0^2 + a_1^2 + a_2^2 + a_3^2$, and the *trace* $t(\alpha) = \alpha + \bar{\alpha} = 2a_0$. As usual we identify a rational number a with the quaternion $a \cdot 1$. Note also the formula $\overline{\alpha\beta} = \bar{\alpha}\bar{\beta}$ holding for all α, $\beta \in \mathbf{H}$. If α is not a rational number then its minimal polynomial is

$$f_\alpha(X) = (X - \alpha)(X - \bar{\alpha}) = X^2 - t(\alpha)X + n(\alpha)$$

Now it is easily seen how quaternions can be used in the questions discussed in the introduction:

(i) The form $x^2 + y^2 + z^2$ is the norm form of quaternions with trace 0.

(ii) If $\mu \in \mathbf{H}$, such that $t(\mu) = 0$, $n(\mu) = m$ (which is equivalent to $\mu^2 = -m$), we have the (embedding) isomorphism ϕ_μ from

$$k := \mathbb{Q}(\sqrt{-m}) \quad \text{to} \quad k_\mu := \mathbb{Q}(\mu) \subset \mathbf{H}$$

which maps $\sqrt{-m}$ on μ.

We note (and use later) the important theorem of Hurwitz's which can be proved by linear algebra.

PROPOSITION 1. If $\mu \in \mathbf{H}$, $\mu \notin \mathbb{Q}$, then the centralizer of k_μ in $\mathbf{H}$ (that is, the set of $\alpha \in \mathbf{H}$, such that $\alpha\mu = \mu\alpha$) is k_μ.

COROLLARY 1. If $\mu, \nu \in \mathbb{H}$ and $\mu^2 = \nu^2 = -m$, then $\{\alpha \in \mathbb{H}; \alpha\mu = \nu\alpha\}$ is a right k_μ-vector space of rank 1 (hence of rank 2 over $\mathbb{Q}$).

b. Some Arithmetical Properties of H

The arithmetics of quaternions is based on the following maximal ring of integral quaternions:

$$\mathbb{E} = \mathbb{Z}i_1 + \mathbb{Z}i_2 + \mathbb{Z}i_3 + \mathbb{Z}\delta, \qquad \delta = (1 + i_1 + i_2 + i_3)/2$$

The simplicity of arithmetics in $\mathbb{E}$ stems from the following facts:

PROPOSITION 2. (i) Every left (right) fractional ideal A of $\mathbb{E}$ can be written $A = \mathbb{E}\alpha$ (resp. $A = \alpha\mathbb{E}$), $\alpha \in \mathbb{H}$, $\alpha \neq 0$. α is determined by A up to a left (resp. right) unit factor.

(ii) The unit group $\mathbb{E}^*$ of $\mathbb{E}$ is generated by $-i_1$, $-i_2$, δ and contains 24 elements.

(i) is proved, for integral ideals, by division with remainder, the norm of which can be made smaller than the norm of the divisor. (ii) is easy, since $\mathbb{E}^* = \{\varepsilon \in \mathbb{E}; n(\varepsilon) = 1\}$.

Now let m be a natural number, $m \equiv 1, 2 \pmod 4$. Then the set of $(x,y,z) \in \mathbb{Z}^3$ such that $x^2 + y^2 + z^2 = m$ is, by $(x,y,z) \to xi_1 + yi_2 + zi_3$, in bijective correspondence with the *set of roots* $R = \{\mu \in \mathbb{E}; \mu^2 = -m\}$, hence card $R = \psi(m)$ (for the definition of $\psi(m)$ see Theorem 1, Introduction). It is a well-known theorem of elementary number theory that R is not empty (cf. [1], Theorem 84; a quaternion proof is available too).

We want to connect the arithmetic of k_μ and $\mathbb{H}$. The maximal order e_μ of k_μ has 1, μ as a $\mathbb{Z}$-basis (note $m \equiv 1, 2 \pmod 4$)), hence $e_\mu \subseteq \mathbb{E}$.

PROPOSITION 3. If a is a fractional ideal of e_μ, then $\mathbb{E}a \cap k_\mu = a$.

Proof. We may assume $a \subseteq e_\mu$. $\mathbb{E}a$ is the $\mathbb{Z}$-submodule of $\mathbb{H}$ generated by all products $\gamma\alpha$, $\gamma \in \mathbb{E}$, $\alpha \in a$; hence it is a left $\mathbb{E}$-ideal. As $1 \in \mathbb{E}$, $a \subseteq k_\mu$, we have $\mathbb{E}a \cap k_\mu \supseteq a$.

On the other hand, a is invertible: there exists a $\mathbb{Z}$-module $a^{-1} \subseteq k_\mu$ such that $a \cdot a^{-1} = e_\mu$. Clearly $(\mathbb{E}a \cap k_\mu)e_\mu = \mathbb{E}a \cap k_\mu$, hence $\mathbb{E}a \cap k_\mu = (\mathbb{E}a \cap k_\mu)a^{-1}a \subseteq (\mathbb{E}aa^{-1} \cap k_\mu a^{-1})a = (\mathbb{E}e_\mu \cap k_\mu)a = (\mathbb{E} \cap k_\mu)a = e_\mu a = a$.

III. CLASS GROUPS AND ROOT BUNDLES

The principal idea is to connect an ideal a in the ideal group $\mathbb{G}$ of k with a specific inner automorphism mapping the root μ onto $\nu \in \mathbb{E}$. Then ν is a root of the same polynomial $X^2 + m$. Thus the finite set of these roots will contain information on ideals, and as it turns out, on ideal classes. The construction is as follows:

If $a \in \mathbb{G}$, put $a_\mu = \phi_\mu(a)$. Then $\mathbb{E}a_\mu$ is a left fractional ideal of $\mathbb{E}$. By Proposition 2, we may choose $\kappa = \kappa(a,\mu) \in \mathbb{H}$ such that

$$\mathbb{E}a_\mu = \mathbb{E}\kappa \tag{0}$$

As κ is determined by a and μ up to left multiplication with units, it is clear that $\nu := \kappa\mu\kappa^{-1}$ is determined up to inner automorphisms induced by units. That $\nu \in \mathbb{E}$ will be shown later. Hence, if for any $\nu \in \mathbb{E}$ with $\nu^2 = -m$ we call the set $B(\nu) = \{\varepsilon\nu\varepsilon^{-1}, \varepsilon \in \mathbb{E}^*\}$ *bundle* of the root ν, we are led to consider the set W of *root bundles*

$$W = \{B(\nu);\ \nu \in \mathbb{E},\ \nu^2 = -m\} \tag{1}$$

Now we define some sort of "pairing" of ideals and root bundles. We look at the map

$$\Delta : \mathbb{G} \times W \to W$$

defined by

$$\Delta(a,B(\mu)) = B(\kappa\mu\kappa^{-1}) \tag{2}$$

where $\kappa = \kappa(a,\mu)$ is chosen such that $\mathbb{E}a_\mu = \mathbb{E}\kappa$. The following properties of Δ are most important.

PROPOSITION 4. (i) Equation (2) actually defines a map $\mathbb{G} \times W \to W$.

(ii) The left kernel of Δ is the group $\mathbb{P}$ of principal ideals in $\mathbb{G}$, that is: $\Delta(a,B(\mu)) = B(\mu)$ is tantamount to a being a principal ideal.

(iii) $W = \{\Delta(a,B(\mu));\ a \in \mathbb{G}\}$.

(iv) If $a,b \in \mathbb{G}$ and μ is a root, then $\Delta(ab,B(\mu)) = \Delta(a,\Delta(b,B(\mu)))$.

Proof. Let κ and $a_\mu = \phi_\mu(a)$ be as described above. Then $\mathbb{E}\kappa\mu\kappa^{-1} = \mathbb{E}a_\mu\mu\kappa^{-1} = \mathbb{E}\mu a_\mu\kappa^{-1} \subseteq \mathbb{E}a_\mu\kappa^{-1} = \mathbb{E}\kappa\kappa^{-1} = \mathbb{E}$, hence $\kappa\mu\kappa^{-1}$ is a root ($\kappa\mu\kappa^{-1} \in \mathbb{E}$). The bundle of $\kappa\mu\kappa^{-1}$ does not depend on which κ we choose in $\mathbb{E}a_\mu = \mathbb{E}\kappa$. We show that the image of Δ depends only on the bundle $B(\mu)$. If $\varepsilon \in \mathbb{E}^*$ and

$\lambda = \varepsilon\mu\varepsilon^{-1}$ we may choose $\kappa(a,\lambda) = \kappa\varepsilon^{-1}$ since $\mathbb{E}a_\lambda = \mathbb{E}\varepsilon a_\mu \varepsilon^{-1} = \mathbb{E}\kappa\varepsilon^{-1}$. This entails

$$\kappa(a,\lambda) \cdot \lambda \cdot \kappa(a,\lambda)^{-1} = \kappa\varepsilon^{-1}\varepsilon\mu\varepsilon^{-1}(\varepsilon\kappa)^{-1} = \kappa\mu\kappa^{-1}$$

as required.

(ii) If $a = e\alpha \in \mathfrak{C}$ is a principal ideal (e the maximal order of k, $\alpha \in k^*$) then $\phi_\mu(a) = e_\mu\beta$, $\beta = \phi_\mu(\alpha)$. We may choose $\kappa = \beta \in k_\mu$ and obtain $\Delta(a,B(\mu)) = B(\beta\mu\beta^{-1}) = B(\mu)$.

To prove the converse we note that, if $\Delta(a,B(\mu)) = B(\mu)$, we may select $\kappa = \kappa(a,\mu)$ such that $\kappa\mu\kappa^{-1} = \mu$. By Proposition 1 we see $\kappa \in k_\mu$, $e_\mu\kappa \subseteq k_\mu$. Hence $\mathbb{E}a_\mu = \mathbb{E}\kappa = \mathbb{E}e_\mu\kappa$ and Proposition 3 implies $a_\mu = e_\mu\kappa$ and a_μ as well as a are principal ideals. Since we know now that Δ depends only on the ideal class, we may restrict attention to integral ideals since every class contains integral ideals. (iv) is a matter of verification. Let

$$\lambda = \kappa(b,\mu) \cdot \mu \cdot \kappa(b,\mu)^{-1}; \quad \text{that is,} \quad B(\lambda) = \Delta(b,B(\mu))$$

Then ϕ_λ is ϕ_μ, followed by the inner automorphism given by $\kappa(b,\mu)$. Hence

$$a_\lambda = \kappa(b,\mu)a_\mu\kappa(b,\mu)^{-1} \quad \text{and}$$

$$\mathbb{E}\kappa(a,\lambda) = \mathbb{E}a_\lambda = \mathbb{E}\kappa(b,\mu)a_\mu\kappa(b,\mu)^{-1}$$

$$= \mathbb{E}b_\mu a_\mu\kappa(b,\mu)^{-1}, \quad \text{that is,} \quad \mathbb{E}\kappa(a,\lambda) \cdot \kappa(b,\mu) = \mathbb{E}(ba)_\mu$$

We may choose $\kappa(ab,\mu) = \kappa(ba,\mu) = \kappa(a,\lambda)\kappa(b,\mu)$, and find

$$\kappa(a,\lambda)\lambda\kappa(a,\lambda)^{-1} = \kappa(ab,\mu)\mu\kappa(ab,\mu)^{-1}$$

as desired.

(iii) This transitivity statement is the most difficult of the proposition; in its proof the hypothesis $m \equiv 1, 2 \bmod 4$ is used. We need some information on the set $T_{\mu,\nu} = \{\lambda \in \mathbb{E};\ \lambda\mu = \nu\lambda\}$, where $\mu,\nu \in \mathbb{E}$ are roots of the polynomial $X^2 + m$. This is obtained in the following proposition.

PROPOSITION 5. For any roots μ, ν there holds $\mathbb{E} \cdot T_{\mu,\nu} = \mathbb{E}$.

Proof. $\mathbb{E} \cdot T_{\mu,\nu}$, the $\mathbb{Z}$-module spanned by all products $\alpha\omega$, $\alpha \in \mathbb{E}$, $\omega \in T_{\mu,\nu}$, is a left ideal of $\mathbb{E}$. Hence it may be written as $\mathbb{E}\rho, \rho \in \mathbb{E}$ (ρ is a common right divisor of all elements of $T_{\mu,\nu}$).

First we show that $2 \nmid n(\rho)$. It is sufficient to see that for some $\omega \in T_{\mu,\nu}$ there holds $2 \nmid n(\omega)$. Note that because of $\nu^2 = \mu^2 = -m$ we have $\alpha\mu + \nu\alpha \in T_{\mu,\nu}$ for every $\alpha \in \mathbb{E}$. Let us suppose $m \equiv 1 \bmod 4$ and write $\mu = x_1 i_1 + x_2 i_2 + x_3 i_3$ and $\nu = y_i i_1 + y_2 i_2 + y_3 i_3$ with x_i, $y_i \in \mathbb{Z}$. Since $m = x_1^2 + x_2^2 + x_3^2 \equiv 1 \bmod 4$, two of the x_i have to be $\equiv 0 \bmod 2$ and one is $\equiv 1 \bmod 2$. By symmetry, we may suppose $x_1 \equiv x_2 \equiv 0 \bmod 2$ and $x_3 \equiv 1 \bmod 2$. As to ν, it is enough to consider two cases, namely

(I) $y_1 \equiv y_2 \equiv 0 \bmod 2$, and $y_3 \equiv 1 \bmod 2$

or (II) $y_1 \equiv y_3 \equiv 0 \bmod 2$, and $y_2 \equiv 1 \bmod 2$

Case (I). If we write $\omega_0 = \mu + \nu$ and $\omega_1 = i_1\mu + \nu i_1$, we see

$$n(\omega_0) = (x_1 + y_1)^2 + (x_2 + y_2)^2 + (x_3 + y_3)^2 \equiv 0 \bmod 4$$

$$n(\omega_1) = (x_1 + y_1)^2 + (x_3 - y_3)^2 + (x_2 - y_2)^2 \equiv 0 \bmod 4$$

and

$$n(\omega_0) - n(\omega_1) = 4x_2y_2 + 4x_3y_3 \equiv 4 \bmod 8$$

Hence one of these norms is $\not\equiv 0 \bmod 8$ and $n(\omega)$ is not divisible by 2 for $\omega = \omega_0/2$ or $= \omega_1/2$. Note: $\omega \in \mathbb{E}$.

Case (II). Here we use $\gamma = (1 + i_2)\mu + \nu(1 + i_2)$. All components of γ, namely, $-(x_2 + y_2)$, $x_1 + x_3 + y_1 - y_3$, $x_2 + y_2$, and $x_3 - x_1 + y_3 + y_1$ are $\equiv 1 \bmod 2$. This implies $\omega = \gamma/2 \in \mathbb{E}$, hence $\omega \in T_{\mu,\nu}$, and $n(\gamma) \equiv 1 + 1 + 1 + 1 \equiv 4 \bmod 8$, yielding $2 \nmid n(\omega)$. A similar discussion settles $m \equiv 2 \bmod 4$.

Note. Proposition 5 breaks down if $m \equiv 3 \bmod 8$. In this case one cannot exclude $2 \mid n(\rho)$. In fact $\mathbb{E}T_{\mu,\nu} = \mathbb{E}(1 + i_1)$ is also possible then. It can be shown that the set of bundles splits into two orbits under the action of the ideal group. This accounts for the doubling of the factor to be used in Gausz's theorem.

We return to Proposition 5. Writing $\mathbb{E} \cdot T_{\mu,\nu} = \mathbb{E}\rho$ we have to show $\rho \in \mathbb{E}^*$, that is, $n(\rho) = 1$. Suppose the contrary: there is a rational prime p dividing $n(\rho)$. We know already $p > 2$.

ρ is a right divisor of $\omega_0 = \mu + \nu$ and $\omega_1 = i_1\mu + \nu i_1$, hence of $i_1(\mu + \nu + i_1(i_1\mu + \nu i_1)) = i_1\nu - \nu i_1$. We conclude $n(\rho)|n(i_1\nu - \nu i_1) = n(2(y_2i_3 - y_3i_2)) = 4(y_2^2 + y_3^2)$, hence $y_2^2 + y_3^2 \equiv 0 \bmod p$. Using i_2, i_3 instead of i_1 we obtain $y_1^2 + y_3^2 \equiv 0 \equiv y_1^2 + y_2^2 \bmod p$. By linear algebra these three congruences have only one solution mod p, namely $y_1 \equiv y_2 \equiv y_3 \equiv 0 \bmod p$. This implies $p^2|y_1^2 + y_2^2 + y_3^2 = m$, a contradiction proving Proposition 5.

Now we are able to prove Proposition 4 (iii): For given roots μ, ν, $T_{\mu,\nu}$ is a $\mathbb{Z}$-module of rank 2 since it is the intersection of the lattice $\mathbb{E}$ with a two-dimensional $\mathbb{Q}$-vector space (cf. corollary of Proposition 1). We may compute a $\mathbb{Z}$-basis $\{\xi,\eta\}$ of $T_{\nu,\mu}$ by linear algebra. Set $a_\mu = e_\mu\xi\bar{\eta} + e_\mu n(\eta)$. Note that $\alpha \in T_{\nu,\mu}$ implies $\bar{\alpha} \in T_{\nu,\mu}$ ($\bar{\alpha}\mu = -\bar{\alpha}\bar{\mu} = \nu\bar{\alpha}$). Hence $\xi\bar{\eta}$ centralizes k_μ. By Proposition 1 we conclude $\xi\bar{\eta} \in k_\mu \cap \mathbb{E} = e_\mu$, finally that a_μ is an ideal of e_μ. By Proposition 5 we find $\mathbb{E}\alpha = \mathbb{E}(e_\mu \cdot \xi\bar{\eta} + e_\mu \cdot \eta\bar{\eta}) = \mathbb{E}\xi\bar{\eta} + \mathbb{E}\eta\bar{\eta} = (\mathbb{E}\xi + \mathbb{E}\eta)\bar{\eta} = \mathbb{E} \cdot T_{\nu,\mu}\bar{\eta} = \mathbb{E}\bar{\eta}$. This shows that we may use $a = \phi_\mu^{-1}(a_\mu)$ and $\bar{\eta} = \kappa(a,\mu)$, and since $\bar{\eta} \in T_{\nu,\mu}$ we have proved $\Delta(a,B(\mu)) = B(\nu)$.

THEOREM 2. Let m be a square free integer $m \equiv 1, 2 \bmod 4$; $\mu \in \mathbb{E}$ such that $\mu^2 = -m$ and e the maximal order of the field $k = \mathbb{Q}(\sqrt{-m})$. Define W as in (1), and the map Δ as in (2). Then $\Delta(a,B(\nu))$ depends only on the ideal class $C = C_a$ of a. The mapping

$$\Pi_C : W \to W$$
$$B(\nu) \to \Delta(a,B(\nu))$$

is a permutation of W, and the mapping $C \to \Pi_C$ is an isomorphism of the ideal class group of k onto a sharply transitive permutation group of W.

Proof. This is, in fact, a corollary of Proposition 4. By (ii) and (iv) we see that $\Delta(a,B(\nu))$ depends only on C_a (use (iv) for a principal ideal b). By (iv), setting $a = b^{-1}$, we see that there is an inverse of the map Π_C, hence Π_C is a permutation. (iv) shows, in addition, that $C \to \Pi_C$ is a group homomorphism into the symmetric group of W.

The image acts transitively on W by (iii). If $\mathbf{P}$ is the principal class, and if $\Pi_C = \Pi_{\mathbf{P}}$, then we obtain for $a \in C$, $b \in \mathbf{P}$ that $\Delta(a,B(\nu)) = \Delta(b,B(\nu)) = B(\nu)$. Hence by (ii) the homomorphism is a monomorphism.

Finally, if $B(\mu)$, $B(\nu) \in W$, there is only one class C, such that $\Pi_C(B(\mu)) = B(\nu)$. For $\Delta(a,B(\mu)) = \Delta(b,B(\mu))$ and $a \in C$ imply $B(\mu) = \Delta(b^{-1},\Delta(b,B(\mu)) = \Delta(b^{-1},\Delta(a,B(\mu)) = \Delta(b^{-1}a,B(\mu))$, hence, by (ii), $b^{-1}a$ is in $\mathbb{P}$, and $b \in C$. Hence the image acts sharply transitive.

As a corollary we obtain Gausz' Theorem 1: By Theorem 2 card (W) = h(m). We have only to count the number of roots belonging to a bundle $B(\mu)$. We operate on the root set R by inner automorphisms induced by $\mathbb{E}^*$. The stabilizer of μ is

$$\{\varepsilon \in \mathbb{E}^*;\ \varepsilon\mu\varepsilon^{-1} = \mu\} = \mathbb{E}^* \cap k_\mu = e^*_\mu = \{\pm 1\}$$

(We used Proposition 1 (iii).) Hence card $B(\mu)$ = index of the stabilizer in $\mathbb{E}^*$ = card $(\mathbb{E}^*)/2 = 24/2 = 12$, and $\psi(m)$ = card (R) = $12 \cdot$ card(W) = 12 h(m).

Finally we point out that we were able to construct (from μ, ν) an ideal a such that $\Delta(a,B(\mu)) = B(\nu)$. If μ is fixed, and ν runs through a full set of representatives of W, the a's form a full set of representatives of the ideal classes of k.

REFERENCES

1. Dickson, L. E. (1939). Modern elementary theory of numbers, The University of Chicago Press

2. Gausz, C. F. (1863). Disquisitiones arithmeticae, Werke, Bd. II, Göttingen

3. Hurwitz, A. (1896). Über die Zahlentheorie der Quaternionen (Nachrichten von der k. Ges. der Wiss, zu Göttingen, Math-phys. Klasse, 313-40)

4. Venkow, V. On the arithmetic of quaternions (Russian). Bulletin de l'Académie des Sciences de l' URSS. Parts I and II (1922), pp. 205-246; Parts III, IV, V (1929), pp. 489-504, 535-562, 607-622.

TWO FACTS CONCERNING RATIONAL 2 x 2 MATRICES LEADING TO INTEGRAL TERNARY FORMS REPRESENTING ZERO

Olga Taussky

California Institute of Technology
Department of Mathematics
Pasadena, California

The ternary forms representing zero are derived from rational numbers which turn up as norms from a quadratic field.

In both cases reported here a routine method also gives the number whose norm turns up. In both cases the noncommutativity of matrices is the essential tool. The method thus gives a characterization, by noncommutativity, of the rational numbers (or more generally, the elements in a field) which are norms from a quadratic extension field.

Let F be a field of characteristic $\neq 2$ and $m \in F, \notin F^2$.

Case I. Let A be a 2×2 matrix with elements in F, and characteristic roots α, $\bar{\alpha}$ in $F(\sqrt{m})$, but not in F. Let S be any matrix with elements in F, with the property

$$S^{-1}AS = A' \qquad \text{(the transpose)} \tag{1}$$

It is known that such an S exists and that in the case under consideration it is necessarily symmetric (cf. [11, 23]). Then

$$-\det S = \text{norm } \lambda, \ \lambda \in F(\sqrt{m}) \qquad \text{(see [10])} \tag{2}$$

Assuming S symmetric (1) implies that A can be expressed in the form

$$A = S_1 S_2, \ S_i = S_i' \tag{3}$$

Consider now the case where A as above is a 2×2 matrix with elements in Z. The special case where both S_i can be chosen with elements in Z was studied recently ([12, 13, 14]). From

$$(S_1S_2)S_1 = S_1(S_2S_1)$$

follows

$$AS_1 = S_1A' \tag{4}$$

or

$$S_1^{-1}AS_1 = A'$$

where

$$-\det S_1 = \text{norm } \lambda, \ \lambda \in Q(\sqrt{m}) \tag{5}$$

In connection with this question two quadratic forms were associated with $A = (a_{ik})$ (see [12, 13, 14]), namely

$$\boxed{F_1 : a_{12}x^2 + (a_{22} - a_{11})xy - a_{21}y^2} \tag{6}$$

This form has the same discriminant as the characteristic polynomial of A, namely

$$(a_{22} - a_{11})^2 + 4a_{12}a_{21} \tag{7}$$

The form is associated with an ideal class in $Z[\sqrt{m}]$ determined by an ideal with Z-basis α_1, α_2 where

$$A\begin{pmatrix}\alpha_1\\ \alpha_2\end{pmatrix} = \alpha\begin{pmatrix}\alpha_1\\ \alpha_2\end{pmatrix} \tag{8}$$

and α_1, α_2 are chosen in $Z[\sqrt{m}]$, see [9]. Equation (8) leads to

$$(a_{11} - \alpha)\alpha_1 + a_{12}\alpha_2 = 0$$

Hence it is easy to check that

$$a_{12}F_1 = (\alpha_1x + \alpha_2y)(\bar{\alpha}_1x + \bar{\alpha}_2y) \tag{9}$$

when - denotes conjugation in $Q(\sqrt{m})$.

The second form, F_2, is obtained in the following way: Consider the ternary diophantine equation which is equivalent to (1)

$$a_{21}x_1 + (a_{22} - a_{11})x_2 - a_{12}x_3 = 0 \tag{10}$$

If it does not vanish identically it has a null space with two base elements. Denote them by

$a_1, a_2, a_3; b_1, b_2, b_3$

Then consider the form

$$F_2 : (a_1a_3 - a_2^2)x^2 + (a_1b_3 - a_2b_2 + a_3b_1)xy + (b_1b_3 - b_2^2)y^2 \tag{11}$$

Its discriminant is the expression (7) divided by the square of g c d $(a_{22} - a_{11}, a_{12}, a_{21})$; see [14].

If $Z[\sqrt{m}]$ is the maximal order in $Q(\sqrt{m})$, then F_2 is associated with the class of the square of the ideal (α_1, α_2) (see [14]). The form F_2 by its origin represents the determinants of all matrices S which are integral, satisfy (1) and are obtained from integral x,y. It is the negative square of F_1 (see also the article by van der Waerden in this book).

The ternary integral form representing zero obtained from (5) is studied further in [19]. There it is linked with a theorem of Rédei and Reichardt concerning ideal classes of order 4 (in the narrow sense) in quadratic fields. On the other hand, ideal classes of order 4 play a role in the integral factorization (3) (see [12]).

Case II. Here we consider two 2 × 2 matrices A, B with elements in a field F of characteristic $\neq 2$, where at least one of them, say A, has characteristic roots not in F, but in a quadratic extension $F(\sqrt{m})$, $m \in F$. Then the following fact holds:

$$-\det(AB - BA) = \text{norm } \lambda, \quad \lambda \in F(\sqrt{m}) \tag{12}$$

This is proved in [15] by a proof giving a routine method for finding λ, (see also [16, 17, 20, 21, 25]). In what follows, several other proofs will be discussed.

Proof by Bender. The matrix A can be transformed to diagonal form

$$\begin{pmatrix} \alpha & \\ & \bar{\alpha} \end{pmatrix}$$

by a similarity of the form

$$S = \begin{pmatrix} \alpha_1 & \bar{\alpha}_1 \\ \alpha_2 & \bar{\alpha}_2 \end{pmatrix}$$

where $-$ means conjugation in the field generated by the eigenvalue α of A and where α_1, α_2 are as in (8). Let β_1, β_2 be defined by trace $(\alpha_i\ \beta_k) = \delta_{ik}$. Then S^{-1} is of the form

$$\begin{pmatrix} \beta_1 & \beta_2 \\ \bar{\beta}_1 & \bar{\beta}_2 \end{pmatrix}$$

Apply this similarity to the F-matrix AB - BA. We obtain

$$\begin{pmatrix} \alpha & \\ & \bar{\alpha} \end{pmatrix} B^{(\alpha)} - B^{(\alpha)} \begin{pmatrix} \alpha & \\ & \bar{\alpha} \end{pmatrix} = \begin{pmatrix} 0 & \lambda \\ \mu & 0 \end{pmatrix}$$

where λ, $\mu \in F(\sqrt{m})$ and $B^{(\alpha)} = S^{-1}BS$. For any F-matrix X we have

$$Y = S^{-1}XS = (y_{ij})$$

with

$$y_{ij} = \Sigma\beta_k^{(i)} x_{k\ell}\alpha_\ell^{(j)} \tag{13}$$

where i, j = 1 or 2 according to whether the element or its conjugate is considered. Hence $y_{12} = \bar{y}_{21}$. For X = AB - BA we obtain $\mu = \bar{\lambda}$.

The other proofs to be discussed lean on the proof of (12) showing that C = AB - BA is 0 or it is nonsingular, if at least one of A, B has characteristic roots not in F (the result (12) can be considered as a more detailed version in the n = 2 case of a theorem observed by Uhlig [24]).

Three proofs are now given of the last fact:

1. Taussky proof. Let C be singular, then C, being a commutator, is similar over F to its companion matrix $\begin{pmatrix} 0 & c \\ 0 & 0 \end{pmatrix}$. Assume $c \neq 0$ and express C in the form

$$\begin{pmatrix} 0 & c \\ 0 & 0 \end{pmatrix} = \begin{pmatrix} x_{11} & x_{12} \\ x_{21} & x_{22} \end{pmatrix}\begin{pmatrix} y_{11} & y_{12} \\ y_{21} & y_{22} \end{pmatrix} - \begin{pmatrix} y_{11} & y_{12} \\ y_{21} & y_{22} \end{pmatrix}\begin{pmatrix} x_{11} & x_{12} \\ x_{21} & x_{22} \end{pmatrix}$$

This leads to the following equations:

$$x_{12}y_{21} - x_{21}y_{12} = 0$$

$$(x_{11} - x_{22})y_{12} - (y_{11} - y_{22})x_{12} = c$$

$$(x_{11} - x_{22})y_{21} - (y_{11} - y_{22})x_{21} = 0$$

In virtue of the first equation and $c \neq 0$ it follows that x_{21}, y_{21} are proportional to x_{12}, y_{12}, and hence x_{21}, $y_{21} = 0$.

This (rather computational) proof shows up another fact: If $C \neq 0$ is singular then both A, B are simultaneously similar to triangular matrices. (This is the n = 2 case of a theorem obtained by Laffey [5] and Guralnick [3]). In fact, if C is already in the form $\begin{pmatrix} 0 & c \\ 0 & 0 \end{pmatrix}$ then A, B are triangular. This fact will be discussed in more detail later.

2. *Kisilevsky's proof*. We may assume tr A = 0. Hence AC + CA = 0;

$$C = \begin{pmatrix} 0 & c \\ 0 & 0 \end{pmatrix} \Rightarrow C^2 = 0 \Rightarrow CAC = 0$$

Hence

$$A = \begin{pmatrix} a_{11} & a_{12} \\ a_{21} & a_{22} \end{pmatrix} \Rightarrow CAC = \begin{pmatrix} 0 & c^2 a_{21} \\ 0 & 0 \end{pmatrix} \Rightarrow a_{21} \text{ or } c = 0$$

3. *Zassenhaus' proof*. This consists in giving an alternative proof of Taussky's observation that C singular implies that A, B are simultaneously similar to triangular matrices (over F); see page 42.

Proof of (12) by Zassenhaus. We may assume tr A = 0. We may further assume A in companion matrix form. Under this assumption we have

$$A^2 = -\det A \cdot I, \quad \text{and} \quad \operatorname{diag} A = 0$$

Denote AB - BA by C. Then

$$AC + CA = 0 \tag{14}$$

This implies

$$C^{-1}AC = -A \tag{15}$$

Let

$$A = \begin{pmatrix} 0 & 1 \\ -d & 0 \end{pmatrix}, \qquad L_0 = \begin{pmatrix} 1 & 0 \\ 0 & -1 \end{pmatrix}$$

then

$$C^{-1}AC = L_0 A L_0 \tag{16}$$

Hence $C = L_0$ times a matrix which commutes with A, i.e., a matrix of the form $aI_2 + bA$, $a,b \in F$. Hence det C = -norm λ, $\lambda \in F(\sqrt{m})$. Zassenhaus further pointed out that the algebra generated by I,A and L_0 displays the ring $F^{2\times 2}$ of 2 × 2 F-matrices as a cyclic algebra. He further reproved (12) via cyclic algebras and extended it to cyclic algebras [25].

Stimulated by this, two independent proofs of (12) via cyclic algebras will now be given.

Proof of (12) by Taussky. Consider the ring $F^{2\times 2}$ as the cyclic algebra with the cyclic subfield generated by I and A. Transform A to diagonal form $\begin{pmatrix}\alpha & \\ & \bar{\alpha}\end{pmatrix}$ via a similarity $S^{-1}AS$ and apply the same similarity to B and $C = AB - BA$. Let $S^{-1}BS$ be equal to

$$\begin{pmatrix} b_{11}(\alpha) & b_{12}(\alpha) \\ b_{21}(\alpha) & b_{22}(\alpha) \end{pmatrix}$$

and $S^{-1}CS = C(\alpha)$. Then

$$\begin{aligned} C(\alpha) &= \begin{pmatrix}\alpha & \\ & \bar{\alpha}\end{pmatrix}\begin{pmatrix} 0 & b_{12}(\alpha) \\ b_{21}(\alpha) & 0 \end{pmatrix} - \begin{pmatrix} 0 & b_{12}(\alpha) \\ b_{21}(\alpha) & 0 \end{pmatrix}\begin{pmatrix}\alpha & \\ & \bar{\alpha}\end{pmatrix} \\ &= \begin{pmatrix} 0 & (\alpha - \bar{\alpha})b_{12}(\alpha) \\ (\bar{\alpha} - \alpha)b_{21}(\alpha) & 0 \end{pmatrix} \end{aligned} \tag{17}$$

for the diagonal of $S^{-1}BS$ commutes with $\begin{pmatrix}\alpha & \\ & \bar{\alpha}\end{pmatrix}$.

Since C is nonsingular $b_{12}(\alpha)b_{21}(\alpha) \neq 0$. Further

$$\begin{pmatrix} 0 & b_{12}(\alpha) \\ b_{21}(\alpha) & 0 \end{pmatrix} \text{ transforms } \begin{pmatrix}\alpha & \\ & \bar{\alpha}\end{pmatrix} \text{ into } \begin{pmatrix}\bar{\alpha} & \\ & \alpha\end{pmatrix}, \text{ and } \begin{pmatrix} 0 & b_{12}(\alpha) \\ b_{21}(\alpha) & 0 \end{pmatrix}^2 = b_{12}(\alpha)b_{21}(\alpha)I$$

The product $b_{12}(\alpha)b_{21}(\alpha) \in F$ via (17) since $\det C \in F$.

We now invoke a fundamental result on cyclic algebras over F. Let the cyclic algebra contain the maximal cyclic subfield generated by an element a and b the invertible element such that $b^{-1}ab = a^{\sigma}$, σ an automorphism of the field generated by a. Then $b^2 \in F$. The algebra is a matrix algebra iff b^2 is the norm of an element from the cyclic field.

Consider then the ring of 2×2 matrices over $F(\alpha)$ obtained from $F^{2\times 2}$ if a similarity transforming A into $\begin{pmatrix}\alpha & \\ & \bar{\alpha}\end{pmatrix}$ is applied to $F^{2\times 2}$. In this algebra I and $\begin{pmatrix}\alpha & \\ & \bar{\alpha}\end{pmatrix}$ generate a quadratic field. The algebra is cyclic and C as in (17) transforms $\begin{pmatrix}\alpha & \\ & \bar{\alpha}\end{pmatrix}$ into $\begin{pmatrix}\bar{\alpha} & \\ & \alpha\end{pmatrix}$. It follows that $b_{12}(\alpha)b_{21}(\alpha)$ is a norm and (12) is proved. This proof shows that the similarity which transforms A into diagonal form transforms any F-matrix B into a matrix whose off-diagonal elements have as product a norm from $F(\alpha)$.

Proof of (12) by Kisilevsky and Taussky. This proof gives a rational and simpler version of the preceding proof. Since $C = AB - BA$ is nonsingular consider $C^{-1}AC$. If $\operatorname{tr} A = 0$ is assumed then this is $-A$ and can be interpreted as A^{σ} in the cyclic algebra with cyclic subfield generated by A and automorphism σ: $A \to -A$. Hence $(AB - BA)^2 = -\det(AB - BA)I$ implies that $-\det(AB - BA)I$ is a norm from the field generated by A to FI. This can be interpreted as $-\det(AB - BA)$ being a norm from $F(\sqrt{m})$ to F.

Further proofs were developed by Zassenhaus and by Estes, the former using quaternions, the latter interpreting $F^{2\times 2}$ as a 4-dimensional quadratic space with bilinear form $1/2 \operatorname{tr} (A \operatorname{adj} B)$. Estes too observed the connections of (12) with cyclic algebras. These proofs are too long to report here.

We return now to the discussion of a singular $AB - BA$. Both matrices A, B have then their characteristic roots in F. As pointed out before, the matrices A, B are then simultaneously similar to triangular matrices over F. Zassenhaus gives the following alternative proof for this.

If $AB - BA$ is singular, the linear system

$$(AB - BA)x = 0$$

has a solution $x \neq 0$. This implies

$$ABx = BAx$$

Further

$$(AB - BA)Bx = (AB^2 - BAB)x = (AB^2 - B^2A)x = 0$$

if also $\operatorname{tr} B = 0$ is assumed which is no restriction. It follows that either $Bx = 0$ or Bx is a multiple of x since $AB - BA$ is $\neq 0$, but nilpotent and has only one characteristic vector. If $Bx = 0$ then $AB - BA$ and B have a common characteristic vector x. The same argument carried out for A leads in all cases to the conclusion that A, B have a common characteristic vector, hence they are similar to triangular matrices simultaneously. (For references concerning simultaneous triangular matrices, see [6].)

APPENDIX

Here some simple facts are reported concerning norms of ideals in quadratic fields defined by the characteristic roots of 2×2 integral matrices. These facts are connected with observations of Ochoa [7].

1. Let $A = (a_{ik})$ be an integral 2×2 matrix with irrational characteristic roots in $Q(\sqrt{m})$. Let S run through all unimodular integral 2×2 matrices. Let a_{12}^S be the element in position (12) in the matrix $S^{-1}AS$. Let S_1, S_2 be any two choices of S. Then $a_{12}^{S_1}/a_{12}^{S_2}$ is norm from $Q(\sqrt{m})$.

Proof. Let $S = \begin{pmatrix} \alpha & \beta \\ \gamma & \delta \end{pmatrix}$ with α, β, γ, $\delta \in Z$ and $\alpha\delta - \beta\gamma = \pm 1$. Then $a_{12}^S = (-a_{21}\beta^2 + (a_{22} - a_{11})\beta\delta + a_{12}\delta^2)$. This quadratic form in β, δ, as remarked earlier (see page 40), corresponds to an ideal associated with the matrix class of A. After multiplication by a_{12} it represents the norms of numbers in this ideal.

2. Let $A = (a_{ik})$ be a matrix as in 1. Assume that the characteristic roots of A generate the maximum order in $Q(\sqrt{m})$. Then

$$a_{12} = \text{norm}\ \mathfrak{a}$$

where $\mathfrak{a}$ is an ideal in the class corresponding to the matrix class of A, hence determined up to a factor $\lambda \in Q(\sqrt{m})$. [This agrees with the result from 1.]

Proof. The assertion follows from the fact that the quadratic form corresponding to $\mathfrak{a}$ is congruent to

$$a_{21}x^2 + (a_{22} - a_{11})xy - a_{12}y^2$$

REFERENCES

1. E. Bender, private communication.

2. D. Estes, Determination of Galois automorphisms of maximal commutative rings of 2×2 matrices, Linear Alg. and Appl. 27 (1979), 225-243.

3. R. Guralnick, A note on pairs of matrices with rank one commutator, Lin. and Multilin. Alg. 8 (1979), 97-99.

4. H. Kisilevsky, The Rédei-Reichardt Theorem--a new proof, this volume.

5. T. Laffey, Simultaneous diagonalization of matrices--low rank cases and the nonderogatory case, Lin. and Multilin. Alg. 6 (1978), 269-305.

6. N. H. McCoy, On the characteristic roots of matrix polynomials, Bull. AMS 42 (1936), 592-600.

7. J. Ochoa, Un modela elemental para la clases de ideales de un anillo algebrico, Revista de la R. Academia de Ciencias Exactas 68 (1974), 771-806.

8. L. Rédei and H. Reichardt, Die durch 4 teilbaren Invarianten der Klassengruppe des quadratischen Zahlkörpers, J. Reine Angew. Math. 170 (1933), 69-74.

9. O. Taussky, On a theorem of Latimer and MacDuffee, Can. J. Math. 1 (1949), 300-302.

10. ———, Ideal Matrices I. Arch. Math. 13 (1962), 275-282.

11. ———, Symmetric matrices and their role in the study of general matrices, Linear Alg. and Appl. 5 (1972), 147-159.

12. ———, The factorization of an integral matrix into a product of two integral symmetric matrices I, Acta Arith. 24 (1973), 151-156.

13. ———, The factorization of an integral matrix into a product of two integral symmetric matrices II, Comm. Pure and Appl. Math. 26 (1973), 847-854.

14. ———, The factorization of an integral matrix into a product of two integral symmetric matrices, Bull. AMS 79 (1973), 956-958.

15. ———, Additive commutators between 2 × 2 integral matrix representations of orders in identical or different quadratic number fields, Bull. AMS 80 (1974), 885-887.

16. ———, Commutators of rational 2 × 2 matrices, Linear Alg. and Appl. 12 (1975), 1-6.

17. ———, Norms from quadratic fields and their relation to noncommuting 2 × 2 matrices II, The principal genus, Houston Math. J. (1977),

18. ———, Norms from quadratic fields and their relation to noncommuting 2 × 2 matrices III, A link between the 4-rank of the ideal groups in 543-546.

19. ———, Norms from quadratic fields and their relation to noncommuting 2 × 2 matrices III, a link between the 4-rank of the ideal groups in $Q(\sqrt{m})$ and $Q(\sqrt{m})$, Math. Zeitschr. 154 (1977), 91-95.

20. ———, Norms from quadratic fields related to noncommuting 2 × 2 matrices, Abstract Kyoto Conference in Algebraic Number Theory, 1976.

21. ———, From cyclic algebras of quadratic fields to central polynomials, J. Austral. Math. Soc. 25 (1978), 503-506.

22. O. Taussky, More on norms from algebraic number fields, commutators and matrices which transform a rational matrix into its transpose, Linear Alg. and Appl. 29 (1980), 456-464.

23. O. Taussky and H. Zassenhaus, On the similarity transformation between a matrix and its transpose. Pac. J. Math. 9 (1959), 893-896.

24. F. Uhlig, A rational canonical pair form for a pair of symmetric matrices over an arbitrary field F with characteristic $\neq 2$ and application to finest simultaneous block diagonalization, Lin. and Multilin. Alg. 8 (1979), 41-67.

25. H. Zassenhaus, A theorem on cyclic algebras, Number Theory and Algebra, Academic Press, 1977, 363-393.

ON THE 1ST AND THE 2ND QUADRATIC FORM*

B. L. van der Waerden

Department of Mathematics
University of Zürich
Zürich, Switzerland

Dear Mrs. Taussky-Todd:

I believe I can now fully answer your question why your 1st and 2nd quadratic form have the same discriminant, making use of the theory of invariants.

I. REVIEW OF BASIC THEOREMS

Firstly I shall review the requisite theorems about invariants of binary and ternary forms.

Notations:

Linear forms $$a_x = \begin{cases} a_1x^1 + a_2x^2 + a_3x^3 & \text{ternary} \\ a_1x^1 + a_2x^2 & \text{binary} \end{cases}$$

Determinants
$$(abc) = a_1b_2c_3 + \cdots - \cdots \quad \text{ternary}$$
$$(ab) = a_1b_2 - a_2b_1 \quad \text{binary}$$

also called "bracket factors"

THEOREM 1. Every invariant of the linear forms a_x, b_x, ... is expressible in terms of bracket factors.

Proved so far only for zero characteristic.

*Two letters of B. L. van der Waerden to Olga Taussky-Todd (translated from the German originals by H. Zassenhaus, Salzburg, September 22, 1973)

N.B. One considers as invariants of forms $f(x)$, $g(x)$, ..., only *homogeneous* expressions in the coefficients. For any linear transformation $x = Tx'$ the invariants are multiplied by a power of det T with exponent depending only on the invariant.

THEOREM 2. All invariants of degree m in the coefficients of a form $f(x)$ can be derived from multilinear invariants of m forms f, g, ... by means of equalizing those m forms afterwards.

Proved also only for zero characteristic.

THEOREM 3. In order to form linear invariants of a form f of degree n it suffices to interpret the form as nth power of a linear form. Thus one sets 'symbolically'

$$f(x) = f_1(x)^n$$

Example. Given the quadratic form

$$g(x) = \Sigma\, g_{ik}x^i x^k$$

then let 'symbolically' $g_{ik} = g_i g_k$, form linear functions of the products $g_i g_k$ and substitute afterwards the g_{ik} for the $g_i g_k$.

This is valid only for zero characteristic.

II. EXAMPLES FOR THEOREMS 1, 2, AND 3

Example 1. Question: How does one form the discriminant of a binary form

$$g(x) = g_{11}x^1x^1 + 2g_{12}x^1x^2 + g_{22}x^2x^2$$

Answer: One writes symbolically $g(x) = a_x^2 = b_x^2$ and forms the bilinear invariant

$$\begin{aligned}(ab)^2 = (a_1b_2 - a_2b_1)^2 &= a_1^2b_2^2 - 2a_1a_2b_1b_2 - a_2^2b_1^2 \\ &= g_{11}g_{22} - 2g_{12}g_{12} + g_{22}g_{11} \\ &= 2(g_{11}g_{22} - g_{12}^2) = 2\Delta\end{aligned}$$

Example 2. Analog: The determinant of a ternary form $f(x) = a_x^2 = b_x^2 = c_x^2$ is $(abc)^2$.

Example 3. Given a ternary quadratic form $f(x) = a_x^2 = b_x^2$ ($f = 0$ defines a conic section K) and a linear form u_x ($u_x = 0$ defines a straight linear). What is the condition that the straight line u be tangent to the conic section K?

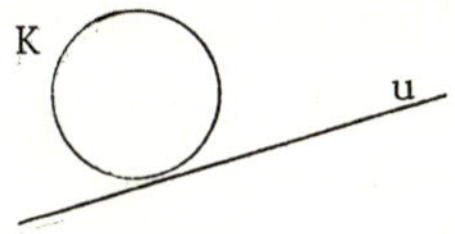

Answer: $(abu)^2 = 0$. (If you don't believe it, please do the calculation!)

III. AN IMPORTANT PRINCIPLE

Notation: $\widehat{yz} = u$ means that

$$\begin{aligned} y^1z^2 - y^2z^1 &= u_3 \\ y^2z^3 - y^3z^2 &= u_1 \\ y^3z^1 - y^1z^3 &= u_2 \end{aligned}$$

with invariance only under the *special linear group* SL(3)

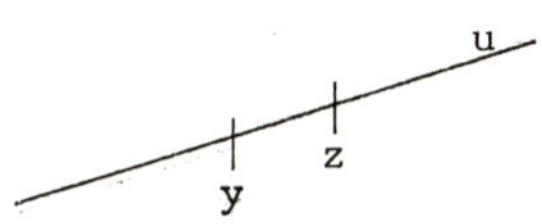

(The straight line u connects the points y and z.)

Additional formula: If $u = \widehat{yz}$ then $(abu) = (ab\widehat{yz}) = a_yb_z - a_zb_y$.

Principle of transference (by Clebsch, I believe).

Given: Ternary forms $f = a_x^n$, $g = c_x^m$, ..., and a linear form u_x.

Geometric realization: Some algebraic curves $f = 0$, $g = 0$, ..., and a straight line u.

Parametric presentation of u:

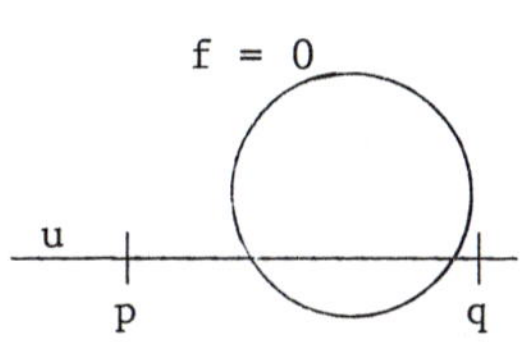

$$x = \xi^1 p + \xi^2 q \qquad \text{i.e., } x^i = \xi^1 p^i + \xi^2 q^i \tag{1}$$

when the points p, q are normed in such a way that $\widehat{pq} = u$.

Substitute (1) in f, g, ...,

$$
\begin{aligned}
f(x) &= f(\xi^1 p + \xi^2 q) = \phi(\xi^1, \xi^2) = \phi \\
g(x) &= \quad \ldots \quad \ldots \quad \ldots \quad = \psi
\end{aligned}
\tag{2}
$$

Now form some invariant I of the forms ϕ, ψ, ...

Question: How to present I as invariant of f, g, ..., and u?

Answer: Let $\phi = \alpha_\xi^n$, $\psi = \gamma_\xi^m$, ..., then I can be expressed in terms of factors like $(\alpha\beta)$, $(\alpha\gamma)$, etc. Now replace $(\alpha\beta)$ by (abu), $(\alpha\gamma)$ by (acu), ..., then one obtains the correct expression for I as invariants of

$$f = a_x^n = b_x^n = \cdots, \qquad g = c_x^m = \cdots \qquad \text{etc.}$$

Proof of the principle of transference. One can indeed proceed as if f really is an nth power of some linear form a_x. Then one is dealing with the invariants of true linear forms a_x, b_x, c_x, ..., and u_x. Now

$$(abu) = (ab\widehat{pq}) = a_p b_q - a_q b_p \tag{3}$$

But $a_x = \alpha_\xi$ according to the definition of the binary form α_ξ, hence in particular for $\xi^1 = 1$, $\xi^2 = 0$, $x = p$, $a_p = \alpha_1$ (similarly, $b_p = \beta_1$, etc.) and in the same way for $x = q$, $\xi^1 = 0$, $\xi^2 = 1$, $q = \alpha_2$ (and $b_q = \beta_2$, etc.). Therefore one can rewrite (3) as $(abu) = \alpha_1\beta_2 - \alpha_2\beta_1 = (\alpha\beta)$. Q.E.D.

Example. Intersect the conic section K, defined by the ternary quadratic form $f(x) = a_x^2 = b_x^2$, with the straight line u; i.e., substitute

$$x = \xi^1 p + \xi^2 q$$

in f(x). Then one obtains the binary form

$$\phi(\xi) = \alpha_\xi^2 = \beta_\xi^2$$

What is its discriminant? Answer:

$$(\alpha\beta)^2 = (abu)^2 \tag{4}$$

IV. O. TAUSSKY'S PROBLEM

After these preparations I shall discuss your problem. The condition AS = SA' leads to your equation

$$a_{21}x + (a_{22} - a_{11})y - a_{12}z = 0 \tag{5}$$

$$\left[\text{N.B.}\quad A = \begin{pmatrix} a_{11} & a_{12} \\ a_{21} & a_{22} \end{pmatrix} \in \mathbb{Z}^{2\times 2},\quad S = \begin{pmatrix} x & y \\ y & z \end{pmatrix}.\right]$$

Write x^1, x^2, x^3 instead of x, y, z and

$$u_1x^1 + u_2x^2 + u_3x^3 = 0 \quad \text{or} \quad u_x = 0 \tag{6}$$

instead of (5), where $u_1 = a_{21}$, $u_2 = a_{22} - a_{11}$, and $u_3 = -a_{12}$. I assume that the greatest common divisor of the u_i's is 1, and forget all about the matrix A. Furthermore, I have the quadratic form $f(x) = \det(S) = xz - y^2 = \Sigma\, g_{ik}x^ix^k$ ($g_{13} = g_{31} = 1/2$, $g_{22} = -1$, all other coefficients zero).

The linear form u_x defines a straight line u, the form f(x) defines a conic section K.

Question: In which way is it possible to express the tangent property of u to K by equating an invariant to zero?

There are three methods available which produce exactly the same invariant in the end. They are:

1st method. The equation of K in terms of line coordinates is

$$(abu)^2 = 0$$

according to II, Example 3.

2nd method. Let the two special solutions

$$p = (p^1, p^2, p^3)$$
$$q = (q^1, q^2, q^3)$$

form a basis of the integral solutions of (6). One can normalize p, q such that $\widehat{pq} = u$. Now substitute again

$$x = \xi^1 p + \xi^2 q$$

in f(x) resulting in a quadratic form

$$\phi(\xi) = \alpha_\xi^2 = \alpha_\xi^2$$

That is your *second form.* Your discriminant (multiplied by 2 as in II, Example 1) is $(\alpha\beta)^2$. According to the principle of transference one has

$$I = (abu)^2$$

One obtains therefore exactly the same invariant I by the 2nd method as one does by the 1st method.

Evaluation of I for the special form $f = xz - y^2$ yields

$$I = \frac{1}{2}(4u_1u_3 - u_2^2)$$

thus

$$2I = -4a_{12}a_{21} - (a_{22} - a_{11})^2$$

3rd method. In place of $\begin{pmatrix} x & y \\ y & z \end{pmatrix}$ I write

$$\begin{pmatrix} s^{11} & s^{12} \\ s^{21} & s^{22} \end{pmatrix}$$

I look for a parametric presentation of the conic section K. It is

$$s^{\alpha\beta} = t^\alpha t^\beta \qquad (\alpha, \beta = 1, 2)$$

(superscripts are clearer because the $s^{\alpha\beta}$ are coordinates of a point and so are the t^α).

I substitute this presentation in the form u_x and obtain

$$\begin{aligned} u_1x + u_2y + u_3z &= u_1s^{11} + u_2s^{12} + u_3s^{22} \\ &= u_1(t^1)^2 + u_2t^1t^2 + u_3(t^2)^2 \end{aligned}$$

This is your *first form.* Your discriminant (multiplied by 2) is equal to

$$2(u_1u_3 - \frac{1}{4}u_2^2) = \frac{1}{2}(4u_1u_3 - u_2^2)$$

which is again the same invariant.

Basically my method of proof is the same as yours, but I believe, now one sees better (with less computation), why it is bound to be this way. What happens is: *Any* evaluation of the condition of contact is bound to result in $4u_1u_3 - u_2^2 = 0$ because that happens to be the condition of contact. Therefore all methods are bound to yield the same invariant I, up to a numerical factor.

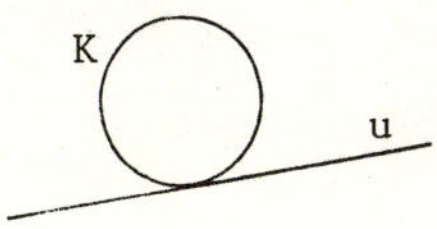

I shall continue to think about the rational equivalence of the first and the second form . . .

B. L. van der Waerden

Zurich, September 29, 1973

Dear Mrs. Taussky:

Now I have worked out in all detail my earlier rash cogitations which had given me the idea that your "first" and "second" form are equivalent.

The outcome is of course that you are right: the two forms are not equivalent. The class of the second form $\phi(\xi)$ multiplied by -1 is precisely the square of the class of the first form.

The parametric presentation of the conic section K was given by

$$\begin{aligned} x^1 &= s^{11} = (s^1)^2 \\ x^2 &= s^{12} = (s^1 s^2) \\ x^3 &= s^{22} = (s^2)^2 \end{aligned} \qquad (1)$$

Now I want to project the points of the conic section from a fixed point x of it on the straight line. There has to be a projective relationship between the parameters t of the points of the conic section and the

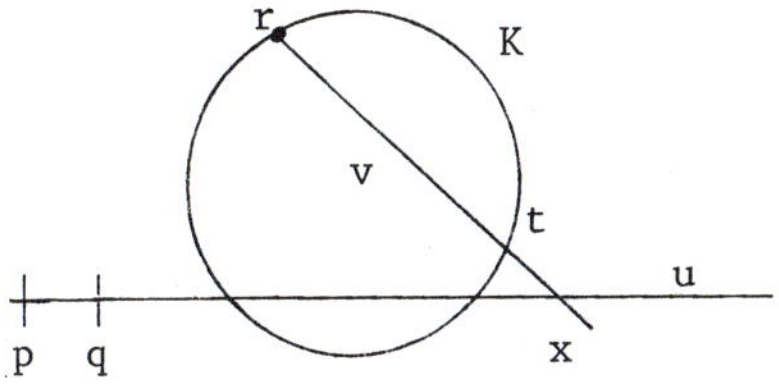

parameters ξ of the points x on u, therefore the ξ's are linear forms of the t's.

Substitution of those linear forms for ξ into the 2nd form $\phi(\xi)$ must result in the first form $f(t)$ up to a factor. Hence the two forms are at least "similar" to each other, i.e., there is a linear transformation transforming the $\phi(\xi)$ up to a factor into $f(t)$. So far my geometric arguments are confirmed by the following computational evaluation.

Let v be the straight line connecting r and t. Substitute the parametric presentation (1) into the equation

$$v_1x^1 + v_2x^2 + v_3x^3 = 0$$

thus

$$v_1(s^1)^2 + v_2s^1 + v_3(s^2)^2 = 0 \tag{2}$$

The zeros of this form are supposed to be r and t. Therefore, the form (2) is equal to

$$(s^1r_2 - s^2r_1)(s^1t_2 - s^2t_1) \tag{3}$$

up to a factor. I can determine the v_i's in such a way that the forms (1) and (2) do not only coincide up to a factor, but that they coincide precisely. Thus

$$\begin{aligned} v_1 &= r_2t_2 \\ v_2 &= -r_1t_2 - r_2t_1 \\ v_3 &= r_1t_1 \end{aligned} \tag{4}$$

In order to find the point of intersection x of v and u = pq, one must substitute the parametric presentation

$$x = \xi^1p + \xi^2q \tag{5}$$

of u into the equation of v (abbreviate $v_x = \Sigma\, v_kx^k$):

$$v_x = \xi^1v_p + \xi^2v_q = 0$$

A solution of this equation is given by

$$\begin{aligned} \xi^1 &= v_q = r_2t_2q^1 - (r_1t_2 + r_2t_1)q^2 + r_1t_1q^3 \\ \xi^2 &= -v_p = -r_2t_2p^1 + (r_1t_2 + r_2t_1)p^2 + r_1t_1p^3 \end{aligned} \tag{6}$$

This is--keeping r fixed--the desired bilinear relation of ξ and t.

The *1st form* f(t) is obtained upon substitution of the parametric presentation (1) of the conic section into the equation of the straight line u:

$$f(t) = u_1t_1^2 + u_2t_1t_2 + u_3t_2^2 \tag{7}$$

The 2nd form $\phi(\xi)$ is obtained upon substitution of the parametric presentation (5) of the straight line into the equation of K:

$$xz - y^2 = x^1x^3 - x^2x^2 = 0$$

The result is:

$$\phi(\xi) = (\xi^1p^1 + \xi^2q^1)(\xi^1p^3 + \xi^2q^3) - (\xi^1p^2 + \xi^2q^2)^2 \qquad (8)$$

Upon substitution of the expressions (6) for ξ^1, ξ^2 one obtains a quadratic form in the t's--by the way also a quadratic form in the r's--and the whole will be symmetric in t and r. The zeros of this form $\Phi(t)$ are the same as the zeros of $f(t)$, namely the parameter values of the points of intersection of u and K. Hence $\Phi(t)$ has to be divisible by $f(t)$. Because of the symmetry in t and r, it also must be divisible by $f(r)$. Computation yields

$$\Phi(t) = \Phi(r,t) = -f(r)f(t) \qquad (9)$$

This formula has satisfied me deeply. It does not only show that the form $\phi(\xi)$ is "similar" to the form $f(t)$, but it shows more: The form $-\phi(\xi)$ is the composite of $f(t)$ with itself in the sense of Gauss. This is because the bilinear substitution (6) carries $-\phi(\xi)$ into $f(r) \cdot f(t)$. In other words: The equivalence class of $-\phi$ is the square of the equivalence class of f.

Isn't that beautiful?

B. L. van der Waerden

CHARACTERISTIC POLYNOMIALS OF SYMMETRIC MATRICES

Alfred Weiss

Department of Mathematics
University of Alberta
Edmonton, Canada

The problem: Given a monic polynomial $f(X)$ with coefficients in a field F decide if $f(X)$ is the characteristic polynomial of some symmetric matrix with entries from F.

The solution of this problem depends on the nature of the field F in a rather involved way, via a certain family of quadratic forms over F. The principal invariants are of an arithmetic type; e.g., the first invariant depends on norms from extension fields of F (see Corollary to Lemma 1). After a discussion of the general case we will present a complete solution when F is a local number field, following Bender [4,5,6].

Initially we assume only that

All roots of $f(X)$ over F lie in a separable extension field of F. (1a)

The characteristic of F is not equal to 2. (1b)

To the monic polynomial $f(X)$ over F we associate an F-algebra $A = A(f)$ as follows. Factor $f(X)$ as

$$f(X) = f_1(X)\cdots f_s(X) \tag{2a}$$

with each $f_i(X)$ irreducible over F, and put

$$A(f) = \prod_{1\le i\le s} K_i \quad \text{with component fields}$$
$$K_i = F[X]/(f_i(X)) \qquad 1 \le i \le s \tag{2b}$$

We write also

$$n = \text{degree } f(X) = \dim_F A$$
$$n_i = \text{degree } f_i(X) = [K_i:F] \qquad 1 \le i \le s \tag{2c}$$

The point of (1a) is that $A(f)$ is a separable (commutative) F-algebra, i.e., the symmetric bilinear form

$$(\nu,\mu) \mapsto \mathrm{tr}_{A/F}(\nu \cdot \mu) \tag{2d}$$

is nondegenerate, so A has also a quadratic structure.

For the purpose of eventual clarity we note that any separable F-algebra A (which is commutative and finite dimensional) can be realized as $A(f)$ for some polynomial $f(X)$ over F, and our problem can be formulated for such algebras A. Namely, letting A act on itself by multiplication defines the regular representation

$$A \to \mathrm{End}_F A \tag{3}$$

which is a natural F-algebra monomorphism. Each F-basis $\underline{\omega}$ of A then represents A as an algebra of matrices over F. If $\underline{\omega}$ can be chosen so that all these matrices are symmetric we say that the F-algebra A is *representable by symmetric matrices.*

Given a separable F-algebra A we now construct a family of quadratic invariants of A. If A^* denotes the group of units of A then there will be an invariant associated to each element of A^*/A^{*2}. For $\lambda \in A^*$ and an F-basis $\underline{\omega} = \{\omega_1,\ldots,\omega_n\}$ of A we put

$$A(\lambda,\underline{\omega}) = [\mathrm{tr}_{A/F}(\lambda \cdot \omega_i\omega_j)]_{1\le i,j\le n} \tag{4a}$$

which is a nonsingular symmetric matrix over F. If $\underline{\omega}'$ is another F-basis then $A(\lambda,\underline{\omega})$, $A(\lambda,\underline{\omega}')$ are congruent over F, i.e., $A(\lambda,\underline{\omega}') = P^t A(\lambda,\underline{\omega})P$ with P an invertible matrix over F. So we put:

$$A(\lambda,A/F) = \text{F-congruence class of } A(\lambda,\underline{\omega}) \tag{4b}$$

Finally, if $\nu \in A^*$ then $A(\lambda\nu^2,\underline{\omega}) = A(\lambda,\nu\underline{\omega})$, where $\nu\underline{\omega} = \{\nu\omega_1,\ldots,\nu\omega_n\}$ is another F-basis of A, so

$$A(\lambda,A/F) \text{ depends only on the image of } \lambda \text{ in } A^*/A^{*2} \tag{4c}$$

and we have formed the basic invariants.

This formulation of our problem is justified by

THEOREM 1. Let $f(X)$ be a monic polynomial over F satisfying (1a) and (1b) and let $A = A(f)$ be the F-algebra (2b) associated to $f(X)$. The following are equivalent:

i. $f(X)$ is the characteristic polynomial of some symmetric F-matrix.

ii. A is representable by symmetric matrices.

iii. The identity matrix I_n is in the F-congruence class $A(\lambda, A/F)$ for some λ in A^*/A^{*2}.

Proof. For $\nu \in A$ and an F-basis $\underline{\omega}$ of A let

$$[\nu]_{\underline{\omega}} = \text{the matrix of multiplication by } \nu \text{ in the basis } \underline{\omega} \tag{5a}$$

in the spirit of (3). If $\underline{\omega}^*$ is the dual basis to $\underline{\omega}$ relative to $tr_{A/F}$, i.e.,

$$tr_{A/F}(\omega_i^* \omega_j) = \delta_{ij} \qquad \text{(Kronecker delta)} \tag{5b}$$

then an easy computation shows that

$$[\nu]_{\underline{\omega}} = [tr_{A/F}(\nu\omega_i^*\omega_j)]_{1\le i,j\le n} \tag{5c}$$

We define the element θ in A by declaring its components

$$\theta_i = X + (f_i(X)) \in K_i \qquad 1 \le i \le s \tag{5d}$$

Letting C_f denote the direct sum of the companion matrices of $f_1(X), \ldots, f_s(X)$, we then have

$$[\theta]_{\underline{\omega}} \text{ is similar to } C_f \text{ over } F \tag{5e}$$

For choose the F-basis $1, \theta_i, \ldots, \theta_i^{n_i - 1}$ for each component field K_i and use these bases together to form a basis of A. In this basis multiplication by θ clearly has matrix C_f.

iii implies ii. By hypothesis A has an F-basis $\underline{\omega}$ and a unit λ so that $A(\lambda,\omega) = I_n$, i.e., $tr_{A/F}(\lambda\omega_i\omega_j) = \delta_{ij}$. This says that $\omega_i^* = \lambda\omega_i$, $1 \le i \le n$, so if ν is any element of A then

$$[\nu]_{\underline{\omega}} = [tr_{A/F}(\nu\lambda\omega_i\omega_j)]$$

by (5c). Since each $[\nu]_{\underline{\omega}}$ is now symmetric, we deduce ii.

ii implies i. Letting $\underline{\omega}$ be the assumed basis in which A is represented by symmetric matrices we have $[\theta]_{\underline{\omega}}$ symmetric. But $[\theta]_{\underline{\omega}}$ has characteristic polynomial f(X) by (5e) so i follows.

i implies iii. We begin by organizing the algebra A a little better. Factor f(X) as

$$f(X) = g_1(X)^{m_1} \cdots g_r(X)^{m_r} \quad \text{over } F$$

where $g_1(X), \ldots, g_r(X)$ are the distinct irreducible factors of f(X). Put $K_i = F[X]/(g_i(X))$ for $1 \leq i \leq r$, and then put

$$A_i = \prod_{m_i \text{ copies}} K_i \qquad 1 \leq i \leq r \tag{6b}$$

Then A_i is a K_i-algebra, with K_i acting "on the diagonal," so if 1_i denotes the identity of A_i then A_i contains the field $K_i 1_i$. As F-algebras we then have

$$A(f) = \prod_{1 \leq i \leq r} A_i \tag{6a}$$

The first purpose of this reorganization is that we can identify the F-subalgebra of A generated by the element θ of (5d) as

$$F(\theta) = \prod_{1 \leq i \leq r} K_i 1_i \tag{6c}$$

For $F(\theta)$ is, by definition, the image of the homomorphism $F[X] \to A$ which sends the indeterminate X to θ. For each $1 \leq i \leq r$ the composite $F[X] \to A \to A_i$ sends X to

$$\theta 1_i = \underbrace{(X + (g_i(X)), \ldots, X + (g_i(X)))}_{m_i \text{ components}} = [X + (g_i(X))]1_i \in K_i 1_i$$

from which it follows that

$$\operatorname{im}(F[X] \to A_i) = K_i 1_i \qquad \ker(F[X] \to A_i) = (g_i(X))$$

for $1 \leq i \leq r$. Since the $g_i(X)$ are co-prime we have $\ker(F[X] \to A) = (g_i(X) \cdots g_r(X))$, by the Chinese Remainder Theorem, and (6c) follows. Considering now A as $F(\theta)$-module, it follows that

$$\operatorname{End}_{F(\theta)} A = \prod_{1 \leq i \leq r} \operatorname{End}_{K_i} A_i \tag{6d}$$

with $\phi = (\phi_i)_{1\le i\le r}$, where the component ϕ_i of ϕ in $\mathrm{End}_{K_i} A_i$ is given by $\phi_i = \phi 1_i = 1_i \phi$.

The second purpose is to examine the natural quadratic structure (2d). For corresponding to (6a), (6b) we have

$$\mathrm{tr}_{A/F}(\lambda) = \sum_{i=1}^{r} \mathrm{tr}_{A_i/F}(\lambda 1_i) \qquad \lambda \in A \tag{6a'}$$

$$\mathrm{tr}_{A_i/F}(\lambda_i) = (\mathrm{tr}_{K_i/F} \circ \mathrm{tr}_{A_i/K_i})(\lambda_i) \qquad \lambda_i \in A_i \tag{6b'}$$

Moreover, we can define for each $\phi \in \mathrm{End}_F A$ an *adjoint* ϕ^* by

$$\mathrm{tr}_{A/F}(\phi^*(\nu) \cdot \mu) = \mathrm{tr}_{A/F}(\nu \cdot \phi(\mu)) \qquad \text{for } \mu, \nu \text{ in } A \tag{7a}$$

And now if ϕ in $\mathrm{End}_{F(\theta)} A$ is written $\phi = (\phi_i)$ as in (6d) we will even have

$$(\phi_i)^* = \phi^* = (\phi_i^*) \tag{7b}$$

Indeed, (6a') implies $(\phi_i)^* = (\phi_i^*)$ where ϕ_i^* is the adjoint of ϕ_i when A_i is considered as F-algebra. And (6b') shows that this ϕ_i^* is unchanged if it is instead interpreted as the adjoint of ϕ_i with A_i considered as K_i-algebra, because ϕ_i is K_i-linear. Thus (7b) is intended in the strongest sense.

We can now return to the proof. Suppose f(X) is the characteristic polynomial of the symmetric matrix S over F. We first show that we may suppose S is also semisimple by using the Jordan decomposition, in F-rational form. For S has a unique decomposition S = D + N in which D is semisimple, N is nilpotent, and DN = ND. Indeed, D, N can be written as d(S), n(S) where d and n are polynomials with coefficients in F (as in Theorem 1.15 of [8], with (1a) replacing the hypothesis that F is perfect). Now D = d(S) is semisimple and symmetric, and the fact that f(X) is the characteristic polynomial of D follows from the uniqueness of the Jordan decomposition (since we can then reduce to considering Jordan canonical forms where it is obvious).

Now S is semisimple with characteristic polynomial f(X), hence it is similar to the matrix C_f of (5e) by the rational canonical form. By (5e) itself it then follows that A has an F-basis $\underline{\omega}$ in which $[\theta]_{\underline{\omega}} = S$ is symmetric, i.e., multiplication by θ has a symmetric matrix in this basis $\underline{\omega}$.

Let $\underline{\omega}^*$ be the dual basis (5b) corresponding to $\underline{\omega}$ and define $\sigma \in \mathrm{End}_F A$ by $\sigma\omega_i = \omega_i^*$ for $1 \le i \le n$. Then $tr_{A/F}(\sigma(\omega_i) \cdot \omega_j) = \delta_{ij} = \delta_{ji} = tr_{A/F}(\omega_i \cdot \sigma(\omega_j))$ and F-bilinearity of the trace form implies that $\sigma^* = \sigma$, in the sense of (7a). Moreover, $[\theta]_{\underline{\omega}}$ is symmetric so by (5c) we have $tr_{A/F}(\theta \cdot \sigma(\omega_i)\omega_j) = tr_{A/F}(\theta \cdot \sigma(\omega_j)\omega_i) = tr_{A/F}(\theta\omega_i \cdot \sigma(\omega_j)) = tr_{A/F}(\sigma^*(\theta\omega_i) \cdot \omega_j) = tr_{A/F}(\sigma(\theta\omega_i) \cdot \omega_j)$ for $1 \le i, j \le n$. By nondegeneracy of the trace form this relation for all j implies that $\theta \cdot \sigma(\omega_i) = \sigma(\theta\omega_i)$ for $1 \le i \le n$. Thus $\sigma \in \mathrm{End}_{F(\theta)}A$. We can therefore write $\sigma = (\sigma_i)_i$ as in (6d) and by (7b) we will have $\sigma_i^* = \sigma_i$, $1 \le i \le r$.

We fix i temporarily and consider σ_i in $\mathrm{End}_{K_i} A_i$. Let $\underline{e}$ be the special K_i-basis of A_i consisting of the algebra idempotents $e_1^{(i)}, \ldots, e_{m_i}^{(i)}$, which correspond to the m_i component fields K_i of (6b). A direct computation shows that the matrix $[\nu]_{\underline{e}}$ of multiplication by ν in the basis $\underline{e}$ is then a diagonal matrix for each ν in A_i. Thus we have

$$tr_{A_i/K_i}\left(\sum_{k=1}^{m_i} \alpha_k e_k^{(i)}\right) = \sum_{k=1}^{m_i} \alpha_k \qquad \text{with } \alpha_k \text{ all in } K_i$$

And from this it follows that if $\phi_i \in \mathrm{End}_{K_i} A_i$ has matrix $[\phi_i]_{\underline{e}}$ in the basis $\underline{e}$, then $[\phi_i^*]_{\underline{e}} = [\phi_i]_{\underline{e}}^t$. Since $\sigma_i^* = \sigma_i$ we see that $[\sigma_i]_{\underline{e}}$ is a symmetric matrix over K_i. By our hypothesis (1b) we deduce the existence of an invertible $\tau_i \in \mathrm{End}_{K_i} A_i$ so that the matrix $[\tau_i^*\sigma_i\tau_i]_{\underline{e}}$ is diagonal. Thus there is a λ_i in A_i so that

$$\tau_i^*\sigma_i\tau_i = \text{multiplication by } \lambda_i$$

Defining λ in A and an invertible $\tau \in \mathrm{End}_{F(\theta)}A$ by their components λ_i, τ_i under (6a), (6d) we have

$$\tau^*\sigma\tau = \text{multiplication by } \lambda$$

because of (7b). But now

$$\begin{aligned} tr_{A/F}(\lambda \cdot (\tau^{-1}\omega_i)(\tau^{-1}\omega_j)) &= tr_{A/F}(\tau^*\sigma\tau(\tau^{-1}\omega_i) \cdot \tau^{-1}\omega_j) = tr_{A/F}(\sigma\omega_i \cdot \omega_j) \\ &= tr_{A/F}(\omega_i^*\omega_j) = \delta_{ij} \qquad \text{for all i, j} \end{aligned}$$

This says $A(\lambda, \tau^{-1}\underline{\omega}) = I_n$ so, since invertibility of λ follows from that of I_n, the theorem is proved.

We compare the matrix class $A(\lambda, A/F)$ with the algebra decomposition (2b) of A, which gives

$$A^*/A^{*2} = \prod_{1\le i\le s} K_i^*/K_i^{*2}$$

$$A(\lambda, A/F) = A(\lambda_1, K_1/F) \oplus \cdots \oplus A(\lambda_s, K_s/F) \qquad \text{(matrix direct sum)} \tag{8a}$$

$$N_{A/F}(\lambda) = \prod_{1\le i\le s} N_{K_i/F}(\lambda_i)$$

where λ_i is the component of λ in K_i, and N is the norm of the regular representation. Now the first invariant of an F-congruence class C of nonsingular symmetric matrices (equivalently: a nondegenerate quadratic form) is the element $\det(C)$ of F^*/F^{*2} which contains the determinant of any matrix in the class. The class $A(1, A/F)$ has special significance since the algebra discriminant $d(A/F)$ by definition satisfies

$$\det A(1, A/F) \equiv d(A/F) \bmod F^{*2} \tag{8b}$$

Moreover, we have $A(\lambda, \underline{\omega}) = [\lambda]_{\underline{\omega}}^t A(1, \underline{\omega})$ by (4a), (5a), hence

$$\det A(\lambda, A/F) \equiv N_{A/F}(\lambda) d(A/F) \bmod F^{*2} \tag{8c}$$

LEMMA 1. Let A be a separable F-algebra as in (2b), (2c) and put $n_* = \Sigma_{i=1}^s (1/2)n_i(n_i - 1)$. The following are equivalent:

a. $\det A(\lambda, A/F) \equiv 1 \bmod F^{*2}$ for some λ in A^*/A^{*2}.

b. $(-1)^{n_*} \equiv N_{A/F}(\mu) \bmod F^{*2}$ for some μ in A^*.

c. $(-1)^{(1/2)n(n-1)} \in N_{A/F}(A^*)F^{*2}$.

Proof. Choose any polynomial $f(X)$ so $A = A(f)$. Then for $1 \le i \le s$ we have $K_i = F(\theta_i)$ where θ_i is a root of $f_i(X)$ and

$$(-1)^{(1/2)n_i(n_i-1)} d(K_i/F) \equiv N_{K_i/F}(f_i'(\theta_i)) \bmod F^{*2}$$

Defining Δ in A^* to have components $\Delta_i = f_i'(\theta_i)$ and combining (8a), (8c) then shows

$$\text{If } \Delta = \mu\Delta^{-1} \text{ then } (-1)^{n_*} \det A(\lambda, A/F) \equiv N_{A/F}(\mu) \bmod F^{*2} \tag{8d}$$

which proves the equivalence of a and b.

Observe that $-1 \in N_{A/F}(A^*)F^{*2}$ if some n_i is odd: for the element of A with -1 in the ith component and 1's elsewhere then has norm -1, by (8a). Thus b and c are both vacuous unless n_i is even for $1 \le i \le s$, when

$$n_* \equiv \sum_{i=1}^{s} \frac{n_i}{2} = \frac{n}{2} = \frac{1}{2} n(n-1) \bmod 2$$

verifying the equivalence of b and c.

COROLLARY. Suppose f(X) is the characteristic polynomial of a symmetric matrix over F. Then

$$(-1)^{(1/2)n(n-1)} \in N_{A/F}(A^*)F^{*2} \qquad \text{with } n = \text{degree } f,\ A = A(f)$$

Proof. The λ of Theorem 1 has $\det A(\lambda, A/F) \equiv \det I_n \equiv 1 \bmod F^{*2}$ so the result follows by Lemma 1.

Remark 1. The corollary is Theorem 2 of [5] where it is proved in a different way, originating in [12]. This other way is more general but could be avoided here because the implication i ⇒ iii of Theorem 1 strengthens results of [1, 7].

Example 1. If F is a finite field (of odd characteristic) then every monic polynomial over F is the characteristic polynomial of a symmetric matrix over F.

Proof. Given f(X) the condition (1a) holds because F is perfect. Since norms from finite extension fields of F are surjective we have $F^* = N_{A/F}(A^*)F^{*2}$, where $A = A(f)$. By Lemma 1 we have $\det A(\lambda, A/F) \equiv 1 \bmod F^{*2}$ for some λ in A^*/A^{*2} and it follows that I_n is in the class $A(\lambda, A/F)$, because the determinant of a nonsingular symmetric $n \times n$ matrix over a finite field classifies it up to congruence (cf. 62 of [10]). Theorem 1 now gives the result claimed.

Example 2. If F is the real number field $\mathbb{R}$ then a real monic polynomial f(X) is the characteristic polynomial of a real symmetric matrix if and only if all roots of f(X) are real.

Proof. It is easy to see that a real symmetric matrix is congruent to I_n if and only if it is positive definite. Combining this with Theorem 1 our assertion is equivalent to: $A(\lambda, A/F)$ is positive definite for some λ in A^*/A^{*2} if and only if A is the "split" algebra $\prod \mathbb{R}$. But suppose A is non-split, i.e., has a component field $\mathbb{C}$, and $A(\lambda, A/F)$ is positive definite. By (8a) and $\mathbb{C}^* = \mathbb{C}^{*2}$ it follows that $A(1, \mathbb{C}/\mathbb{R})$ is positive definite, which is false (if $\underline{i}$ is the basis $\{1,i\}$ then $A(1,\underline{i}) = \begin{bmatrix} 2 & 0 \\ 0 & -2 \end{bmatrix}$). The converse is trivial.

We turn finally to the case of local number fields, i.e., from now on F is a finite extension field of the rational p-adic field $\mathbb{Q}_p$, for some prime number p. We begin by recalling the classification of nonsingular symmetric matrices over F up to congruence (i.e., of nondegenerate quadratic forms up to isometry). The Hilbert symbol $\langle , \rangle_F$ is a nondegenerate bilinear pairing $F^*/F^{*2} \times F^*/F^{*2} \to \{-1,1\}$ which satisfies

$$\langle \alpha, \beta \rangle_F = \langle \beta, \alpha \rangle_F \qquad \alpha,\ \beta \in F^* \tag{9a}$$

The Hasse symbol c_F associates to each congruence class of nonsingular symmetric matrices A over F a number $c_F(A) \in \{-1,1\}$ subject to

$$c_F(A \oplus B) = c_F(A)c_F(B)\langle \det A, \det B \rangle_F \tag{9b}$$

$$c_F(\alpha A) = \langle \alpha, (-1)^{(1/2)n(n+1)}(\det A)^{n+1} \rangle_F c_F(A) \tag{9b'}$$

where $\alpha \in F^*$ and A is $n \times n$

Indeed, c_F is completely determined by these properties together with $c_F([1]) = 1$, where [1] is the 1×1 matrix. The classification:

Nonsingular symmetric $n \times n$ matrices A, B over F are congruent if and only if $\det A \equiv \det B \bmod F^{*2}$ and $c_F(A) = c_F(B)$ (9c)

See Sec. 63 of [10].

We also need the Lifting Formula of Bender [3]: If K/F is a finite extension (of local number fields) and we write $A(\lambda) = A(\lambda, K/F)$ for $\lambda \in K^*$ (the matrix class of (4b)), then we have

$$\langle \mu, \nu \rangle_K = c_F(A(1))c_F(A(\mu))c_F(A(\nu))c_F(A(\mu\nu))\langle N_{K/F}\mu, N_{K/F}\nu \rangle_F \tag{9d}$$

$$\mu,\ \nu \in K^*$$

and also

$$\langle \alpha, \nu \rangle_K = \langle \alpha, N_{K/F}(\nu) \rangle_F \qquad \alpha \in F^*,\ \nu \in K^* \tag{9d'}$$

For local number fields our problem is answered by:

THEOREM 2. (Bender [4]) Let F be a finite extension of $\mathbb{Q}_p$. A monic polynomial f(X) of degree n over F is the characteristic polynomial of a symmetric matrix over F if and only if

a. $(-1)^{(1/2)n(n-1)} \in N_{A/F}(A^*)F^{*2}$, where $A = A(f)$, and

b. if $F = \mathbb{Q}_2$ and f(X) is irreducible of degree 4 over $\mathbb{Q}_2$ then the field $K = \mathbb{Q}_2[X]/(f(X))$ contains a quadratic extension field of $\mathbb{Q}_2$.

Proof. We have $\det I_n = 1$, $c_F(I_n) = 1$ as follows from (9b) by induction. Combining Theorem 1 with the classification (9c) we find that f(X) being the characteristic polynomial of a symmetric matrix over F is equivalent to the existence of a λ in A^*/A^{*2} with $\det A(\lambda) \equiv 1 \bmod F^{*2}$ and $c_F(A(\lambda)) = +1$, where $A = A(f)$ and $A(\lambda) = A(\lambda, A/F)$. By Lemma 1 existence of λ in A^*/A^{*2} with $\det A(\lambda) \equiv 1 \bmod F^{*2}$ is equivalent to condition a of the theorem. Thus we see that Theorem 2 is equivalent to the following statement, which better describes the flavor of the proof.

THEOREM 2*. Let A be a (commutative, finite-dimensional) separable algebra over the local number field F, and write $A(\lambda) = A(\lambda, A/F)$. The following are equivalent:

a. There exist λ in A^*/A^{*2} with $\det A(\lambda) \equiv 1 \bmod F^{*2}$ but all such λ have $c_F(A(\lambda)) = -1$.

b. $F = \mathbb{Q}_2$ and A is a field K so that the field extension $K/\mathbb{Q}_2$ has degree 4 and no proper intermediate field.

Proof. We begin with a useful observation. Suppose that λ in A^*/A^{*2} has $\det A(\lambda) \equiv 1 \bmod F^{*2}$ and that $n = \dim_F A$ is *even*. Then for every $\alpha \in F^*$ we have $A(\alpha\lambda) = \alpha \cdot A(\lambda)$, hence

$$\begin{aligned} &\det A(\alpha\lambda) \equiv 1 \bmod F^{*2} \quad \text{and} \\ &c_F(A(\alpha\lambda)) = \langle \alpha, (-1)^{(1/2)n(n+1)} \rangle_F c_F(A(\lambda)) \end{aligned} \tag{10a}$$

follow from (9b').

a implies b. Let A satisfy condition a and adopt again the notation (2b), (2c) (with the advantage that we are free to choose the polynomial

$f(X)$ so that $A(f) = A$ according to our needs). From (10a) follows

$$\text{If } n \equiv 2 \bmod 4 \text{ then } -1 \in F^{*2} \tag{10a'}$$

since now $(-1)^{(1/2)n(n+1)} = -1$ and since the Hilbert symbol is nondegenerate.

Suppose $p \neq 2$: In this case we construct λ violating condition a by looking at Lemma 1 more closely. By a the existence of λ with $\det A(\lambda) \equiv 1 \bmod F^{*2}$ is assured, so by Lemma 1 there exists μ with $(-1)^{n_*} \equiv N_{A/F}(\mu) \bmod F^{*2}$--the point is that μ can now be chosen so that $\mu_i = \pm 1$ for $1 \leq i \leq s$. As in the second paragraph of the proof of Lemma 1, this is clear if some n_i is odd; if every n_i is even then, as before, $n_* \equiv (1/2)n(n-1) \bmod 2$, so we can now take $\mu = 1_A$ by invoking (10a') in case $n \equiv 2 \bmod 4$. Now let Λ be the ring of integral elements of F. In each component field K_i of A we can find an element θ_i so that the ring of integral elements of K_i is just $\Lambda[\theta_i]$, i.e., $1, \theta_i, \ldots, \theta_i^{n_i-1}$ is a "minimal basis" of K_i/F (cf. page 59 of [9]). Letting $f_i(X)$ be the defining polynomial of θ_i over F and putting $f(X) = f_1(X)\cdots f_s(X)$ we have $A = A(f)$ and can now define Δ and $\lambda = \mu\Delta^{-1}$ as in (8d). Let $\underline{\theta}$ be the F basis of A which consists of the minimal bases just constructed in each component field and consider the symmetric matrix $A(\lambda,\underline{\theta})$. Repeating the calculations (8a), (8b), (8c), (8d) with matrix $A(\lambda,\underline{\theta})$ instead of matrix classes we get $\det A(\lambda,\underline{\theta}) = \Pi_{1\leq i\leq s} \det A(\lambda_i,\underline{\theta}_i) = \Pi_{1\leq i\leq s} N_{K_i/F}(\mu_i/\Delta_i)d(\Lambda[\theta_i]/\Lambda) = \Pi_{1\leq i\leq s} N_{K_i/F}(\mu_i/f_i'(\theta_i)) \cdot (-1)^{(1/2)n_i(n_i-1)} N_{K_i/F}(f_i'(\theta_i)) = N_{A/F}(\mu)(-1)^{n_*} = 1$, not just a congruence mod F^{*2}. Moreover, a typical entry of $A(\lambda,\underline{\theta})$ is $Tr_{A/F}(\lambda_i\theta_i^m) = Tr_{K_i/F}(\mu_i\theta_i^m/f_i'(\theta_i))$, for some i and m, so $A(\lambda,\underline{\theta})$ has entries in Λ, because $\mu_i\theta_i^m$ is in $\Lambda[\theta_i]$ and $1/f_i'(\theta_i)$ is in the complementary module to $\Lambda[\theta_i]$ (by p. 59 of [9], again). But by $p \neq 2$ integral symmetric matrices with determinant 1 are congruent to I_n (by Sec. 92 of [10]). This is because 2 is a unit in Λ and because $\langle\varepsilon,\eta\rangle_F = 1$ when ε, η are units of Λ. Thus $c_F(A(\lambda)) = 1$, contradicting a.

Therefore we may suppose F is a finite extension of $\mathbb{Q}_2$ from now on. In this case the main argument is

LEMMA 2. Let F be a finite extension of $\mathbb{Q}_2$ and let A be a separable F-algebra, $n = \dim_F A$. Suppose there exists a λ in A^*/A^{*2} so that $c_F(A(\lambda\nu)) = c_F(A(\lambda))$ for all ν in A^*/A^{*2} which satisfy $N_{A/F}(\nu) \equiv 1 \bmod F^{*2}$. Then one of the following holds:

a. $n \leq 2$.

b. $F = \mathbb{Q}_2$ and A is a field K so that the field extension $K/\mathbb{Q}_2$ has degree 4 and no proper intermediate field.

Proof. Denote by X (respectively Y) the kernel (resp. cokernel) of the homomorphism $N_{A/F}$: $A^*/A^{*2} \to F^*/F^{*2}$, so obtaining the exact sequence

$$0 \to X \to A^*/A^{*2} \to F^*/F^{*2} \to Y \to 0 \tag{10b}$$

which we can view as an exact sequence of vector spaces over the field $\mathbf{F}_2$. If $K/\mathbb{Q}_2$ is a finite extension then $\dim_{\mathbf{F}_2} K^*/K^{*2} = [K:\mathbb{Q}_2] + 2$, so from (8a) we get

$$\dim A^*/A^{*2} = dn + 2s \qquad \text{where } d = [F:\mathbb{Q}_2] \tag{10c}$$

and then from (10b) it follows that

$$\dim X = d(n - 1) + 2(s - 1) + y \qquad \text{where } y = \dim Y \tag{10c'}$$

We can associate to A a "Hilbert symbol" $\langle,\rangle_A$ by defining

$$\langle\mu,\nu\rangle_A = \prod_{1\leq i\leq s} \langle\mu_i,\nu_i\rangle_{K_i} \quad \text{on } A^*/A^{*2}$$

in accordance with the decomposition (8a). It is immediately clear that $\langle,\rangle_A$ is a nondegenerate bilinear form on the vector space A^*/A^{*2}. Moreover, we have

$$\langle\mu,\nu\rangle_A = c_F(A(1))c_F(A(\mu))c_F(A(\nu))c_F(A(\mu\nu))\langle N_{A/F}\mu,N_{A/F}\nu\rangle_F \tag{10d}$$

for all μ, ν in A^*/A^{*2}, where $A(\lambda) = A(\lambda,A/F)$. This is proved by induction on s, the case s = 1 being the Lifting Formula (9d). To prove the general case, one splits A into component algebras $A_1 \times A_2$ when there is a corresponding splitting (8a) for $A(\lambda)$ which allows us to express the Hasse symbols on the right side of (10d) in terms of corresponding Hasse symbols for A_1, A_2 by (9b). Applying the induction hypothesis and the obvious fact that the "Hilbert symbol" of A is the product of the "Hilbert symbols" for A_1 and A_2, the right side of (10d) becomes $\langle\mu,\nu\rangle_A$ times a product of various Hilbert symbols $\langle,\rangle_F$. Then a tedious computation using (8c), the bilinearity of $\langle,\rangle_F$ and (9a) shows that this product is 1, establishing (10d).

Moreover, for arbitrary μ, ν, λ in A^*/A^{*2}, if we write $\langle\mu,\nu\rangle_A = \langle\mu,\nu\rangle_A\langle\mu,\lambda\nu\rangle_A$ and expand the right side by (10d) we get the more symmetric

$$\langle\mu,\nu\rangle_A = c_F(A(\lambda))c_F(A(\lambda\mu))c_F(A(\lambda\nu))c_F(A(\lambda\mu\nu))\langle N_{A/F}\mu,N_{A/F}\nu\rangle_F \qquad (10d')$$

We can now prove Lemma 2, for if λ is as in its statement then (10d') and the hypothesis immediately imply that $\langle\mu,\nu\rangle_A = 1$ for all μ, ν in X. Thus

$$X \subseteq X^{\perp} \qquad (10e)$$

where $\perp$ is taken with respect to the bilinear form $\langle,\rangle_A$. Since $\langle,\rangle_A$ is nondegenerate, it follows that

$$\dim X \leq \dim X^{\perp} = \dim A^*/A^{*2} - \dim X$$

and by (10c) and (10c') we find that

$$d(n - 2) + 2(s - 1) + 2y \leq 2 \qquad (10e')$$

Since we may assume $n \geq 3$ every term on the left side of (10e') is ≥ 0. In particular, from $d \geq 1$ it follows that $n \leq 4$. Thus $n = 3$ or 4.

Suppose $n = 3$. From (10e') we have $s = 1$ and $y = 0$; in particular, A is a field K. Since $N_{K/F}(\alpha) = \alpha^3$ for $\alpha \in F^*$ the natural map $F^*/F^{*2} \to K^*/K^{*2}$ can be considered as an identification and also $X \cap F^*/F^{*2} = 1$. By $y = 0$ the exact sequence (10b) implies that $K^*/K^{*2} \cong X \oplus F^*/F^{*2}$. However, (9d') shows that $\langle\alpha,\nu\rangle_K = 1$ for $\alpha \in F^*/F^{*2}$, $\nu \in X$, i.e., $F^*/F^{*2} \subseteq X^{\perp}$ which combines with (10e) and our previous conclusion to show $K^*/K^{*2} \subseteq X^{\perp}$. By nondegeneracy of $\langle,\rangle_K$ we deduce $X = 1$ while by (10c') we have $\dim X = 2d > 0$, a contradiction.

Suppose $n = 4$. From (10e') we now have $d = 1$, $s = 1$, $y = 0$, hence $F = \mathbb{Q}_2$ and A is a field K. Moreover, if $K/\mathbb{Q}_2$ contains a quadratic extension $L/\mathbb{Q}_2$ then $y = 0$ is contradicted: for $\mathbb{Q}_2^*/N_{L/\mathbb{Q}_2}(L^*)$ has order 2 by local class field theory while $N_{K/\mathbb{Q}_2} = N_{L/\mathbb{Q}_2} \circ N_{K/L}$ implies $N_{K/\mathbb{Q}_2}(K^*)\mathbb{Q}_2^{*2} \subseteq N_{L/\mathbb{Q}_2}(L^*)$, i.e., Y has $\mathbb{Q}_2^*/N_{L/\mathbb{Q}_2}(L^*)$ as a homomorphic image. Thus Lemma 2 is proved.

We can now return to the proof of Theorem 2*, and because the desired conclusion b is also case b of Lemma 2, the application of Lemma 2 shows that $n \leq 2$.

Suppose $n = 1$. Then $\lambda = 1$ clearly violates condition a of Theorem 2*.

Suppose $n = 2$. By (10a') we have $i^2 = -1$ for some i in F. Writing $A = A(f)$ with $f(X) = X^2 - \alpha$ for some $\alpha \in F^*$ (by Kummer theory if A is a field, and with $\alpha = 1$ if not), we can use $\lambda = 1/f'(\theta) = 1/2\theta$ and the F-basis $\underline{\theta} = \{1,\theta\}$ to construct the matrix

$$A(\lambda,\underline{\theta}) = \begin{bmatrix} tr_{A/F}\left(\frac{1}{2\theta}\right) & tr_{A/F}\left(\frac{1}{2\theta}\cdot\theta\right) \\ tr_{A/F}\left(\frac{1}{2\theta}\cdot\theta\right) & tr_{A/F}\left(\frac{1}{2\theta}\cdot\theta^2\right) \end{bmatrix} = \begin{bmatrix} 0 & 1 \\ 1 & 0 \end{bmatrix}$$

Then $\det A(\lambda) = -1 \equiv 1 \bmod F^{*2}$ and

$$\begin{bmatrix} 1 & i \\ 1/2 & -i/2 \end{bmatrix}^t \begin{bmatrix} 0 & 1 \\ 1 & 0 \end{bmatrix} \begin{bmatrix} 1 & i \\ 1/2 & -i/2 \end{bmatrix} = \begin{bmatrix} 1 & 0 \\ 0 & 1 \end{bmatrix}$$

shows $c_F(A(\lambda)) = 1$ in contradiction to condition a. This proves one direction of Theorem 2*.

b implies a. In the notation of the proof of Lemma 2 we clearly have $n = 4$, $s = 1$, $d = 1$, while $y = 0$ follows from local class field theory; in fact we have $N_{K/\mathbb{Q}_2}(K^*) = \mathbb{Q}_2^*$ since $\mathbb{Q}_2$ is the largest abelian extension of $\mathbb{Q}_2$ which is contained in K (by p. 180 of [11]). By (10c), (10c') it follows that $\dim K^*/K^{*2} = 6$, $\dim X = 3 = \dim \mathbb{Q}_2^*/\mathbb{Q}_2^{*2}$.

Since $K/\mathbb{Q}_2$ has no quadratic subfield Kummer theory implies that the inclusion induces a monomorphism $\mathbb{Q}_2^*/\mathbb{Q}_2^{*2} \to K^*/K^{*2}$. By $N_{K/\mathbb{Q}_2}(\alpha) = \alpha^4 \equiv 1 \bmod \mathbb{Q}_2^{*2}$ for $\alpha \in \mathbb{Q}_2^*$ we have $\mathbb{Q}_2^*/\mathbb{Q}_2^{*2} \subseteq X$ so by comparing $\mathbb{F}_2$-dimensions we deduce $X = \mathbb{Q}_2^*/\mathbb{Q}_2^{*2}$.

Choosing θ so $K = \mathbb{Q}_2(\theta)$ and letting $f(X)$ be the defining polynomial of θ over $\mathbb{Q}_2$ the procedure (8d) of Lemma 1 assures us that putting $\lambda_0 = 1/f'(\theta)$ will yield $\det A(\lambda_0) \equiv 1 \bmod \mathbb{Q}_2^{*2}$. This λ_0 satisfies the existence requirement of condition a, and by the previous paragraph it follows that the λ with $\det A(\lambda) \equiv 1 \bmod \mathbb{Q}_2^{*2}$ all satisfy $\lambda = \alpha\lambda_0$ for some α in $\mathbb{Q}_2^*/\mathbb{Q}_2^{*2}$. Since $c_{\mathbb{Q}_2}(A(\alpha\lambda_0)) = c_{\mathbb{Q}_2}(A(\lambda_0))$ for α in $\mathbb{Q}_2^*/\mathbb{Q}_2^{*2}$ follows from (10a) it remains only to show that $c_{\mathbb{Q}_2}(A(\lambda_0)) = -1$.

Were this false then, by (9c), I_4 would be in the $\mathbb{Q}_2$ congruence class of $A(\lambda_0)$. However the quadratic form $\nu \to tr_{K/\mathbb{Q}_2}(\lambda_0\nu^2)$ corresponding to $A(\lambda_0)$, represents 0 over $\mathbb{Q}_2$ since $tr_{K/\mathbb{Q}_2}(1^2/f'(\theta)) = 0$ (by p. 58 of [9]). Thus it suffices to show that the quadratic form I_4 does *not* represent 0 over $\mathbb{Q}_2$.

So suppose $x_1^2 + x_2^2 + x_3^2 + x_4^2 = 0$ with x_i in $\mathbb{Q}_2$ not all 0. By multiplying by a suitable power of 2 we may suppose the x_i are all in $\mathbb{Z}_2$, the ring of 2-adic integers, and not all x_i are $\equiv 0 \bmod 2\mathbb{Z}_2$. By symmetry we may assume $x_4 \equiv 1 \bmod 2\mathbb{Z}_2$ hence $x_4^2 \equiv 1 \bmod 8\mathbb{Z}_2$, and then by $\mathbb{Z}/8\mathbb{Z} \cong \mathbb{Z}_2/8\mathbb{Z}_2$ it follows that there exist x_1, x_2, x_3 in $\mathbb{Z}$ solving the congruence

$$x_1^2 + x_2^2 + x_3^2 + 1 \equiv 0 \bmod 8$$

Since this is readily verified to be false, the proof of Theorem 2* is now complete.

Remark 2. That the exceptional case b of Theorem 2* is not vacuous is shown by the $\mathbb{Q}_2$-polynomial

$$f(X) = X^4 - 2X^3 - 2X^2 + 2$$

For if $K = \mathbb{Q}_2[X]/(f(X))$ and L = normal closure of $K/\mathbb{Q}_2$ then $\mathrm{gal}(L/\mathbb{Q}_2) \cong A_4$, the alternating group, by p. 20 of [13]. Then by $[K:\mathbb{Q}_2] = 4$ and the structure of A_4 it follows that $\mathrm{gal}(L/K)$ is a maximal subgroup of $\mathrm{gal}(L/\mathbb{Q}_2)$ verifying b.

REFERENCES

1. Bender, E. A., Classes of matrices over an integral domain, Illinois J. Math. 11 (1967), 697-702.

2. Bender, E. A., Characteristic polynomials of symmetric matrices, Pacific J. Math. 25 (1968), 433-441.

3. Bender, E. A., A lifting formula for the Hilbert symbol, Proc. Amer. Math. Soc. 40 (1973), 63-65.

4. Bender, E. A., and Herzberg, N. P., Symmetric matrices, characteristic polynomials, and Hilbert symbols over local number fields, Bull. Amer. Math. Soc. 79 (1973), 518-520.

5. Bender, E. A., Characteristic polynomials of symmetric matrices II: local number fields, Linear and Multilinear Algebra 2 (1974), 55-63.

6. Bender, E. A., and Herzberg, N. P., Characteristic polynomials of symmetric matrices III: some counterexamples, Linear and Multilinear Algebra 2 (1974), 173-178.

7. Faddeev, D. K., On the characteristic equations of rational symmetric matrices, Dokl. Akad. Nauk. SSR, Vol. 58 (1947), 753-754.

8. Hochschild, G., Affine Algebraic Groups, Holden-Day Inc., 1971.

9. Lang, S., Algebraic Number Theory, Addison-Wesley, 1970.

10. O'Meara, O. T., Introduction to Quadratic Forms, Springer-Verlag, 1971.

11. Serre, J. P., Corps Locaux, Hermann, Paris, 1962.

12. Taussky, O., Ideal matrices I, Arch. Math. 13 (1962), 275-282.

13. Weil, A., Exercices dyadiques, Inventiones Math. 27 (1974), 1-22.

GAUSS' THEORY OF TERNARY QUADRATIC FORMS, AN EXAMPLE OF THE THEORY OF HOMOGENEOUS FORMS IN MANY VARIABLES, WITH APPLICATIONS

Hans J. Zassenhaus

Department of Mathematics
Ohio State University
Columbus, Ohio

INTRODUCTION

Let us quote Gauss, p. 292, Sec. 266

> Thus far we have restricted our discussion to functions of the second degree with *two* unknowns and there was no need to give them a special name. But manifestly this argument is only one section of the general treatise concerning *rational algebraic functions which are integral and homogeneous in many unknowns and many dimensions*. Such functions relative to the number of dimensions can properly be divided into *forms of the second, third, fourth degree, etc.* and relative to the number of unknowns into *binary, ternary, quaternary, etc. forms*. Thus the forms we have been considering can be called simply *binary forms of the second degree*. But functions like
>
> $$Ax^2 + 2Bxy + Cy^2 + 2Dxz + 2Eyz + Fz^2$$
>
> where (A, B, C, D, E, F are integers) are called *ternary forms of the second degree* and so forth. We have devoted the present section to the treatment of binary forms of the second degree. But there are many beautiful truths concerning these forms which are properly considered in the theory of ternary forms of the second degree. We will therefore make a brief digression into this theory and will especially treat of those elements which are necessary to complete the theory of binary forms, hoping thereby to please geometers who would be disappointed if we ignore them or treated them in a less natural manner. We must, however, reserve a more exact treatment of this important subject for another occasion because its usefulness far exceeds the limits of this work and because, hopefully, we will be able to enrich the discussion by more profound insights later on. At this time we will completely exclude from the discussion

> quaternary, quinary, etc. forms and all forms of higher degrees.* It is sufficient to draw this broad field to the attention of geometers. There is ample material for the exercise of their genius, and transcendental arithmetic will surely benefit by their efforts.

Let us first present here the principles of the theory of n-ary forms developed by Gauss (Sec. 283) in modern language.

Second, let us outline the principles of a general theory as anticipated by Gauss.

NOTATION

Gauss considers integral quadratic forms, i.e., homogeneous polynomials

$$f(x_1,\dots,x_n) = \sum_{i,k=1}^{n} a_{ik}x_ix_k \tag{1a}$$

of the second degree with rational integral coefficients a_{ik} subject only to the natural symmetry conditions

$$a_{ik} = a_{ki} \tag{1b}$$

For brevity's sake call them simply *n-ary forms* or *forms*.

The data of a given form f of degree n are assembled into a quadratic matrix

$$A = (a_{ik}) = A_f \tag{1c}$$

of degree n characterized by the symmetry condition

$$A = A^T \in \mathbb{Z}^{n\times n} \tag{1d}$$

This is to say, for any matrix $A = (a_{ik})$ satisfying (1d) there is defined the form (1a). And if f, g are forms of n variables satisfying the conditions $f(\xi) = g(\xi)$ for all n-rows ξ over $\mathbb{Z}$, then it follows that $A_f = A_g$. Indeed, the value of the coefficient a_{ik} is derived from the equation

$$2a_{ik} = f\left(e_{1i}^{1\times n} + e_{1k}^{1\times n}\right) - f\left(e_{1i}^{1\times n}\right) - f\left(e_{1k}^{1\times n}\right) \tag{1e}$$

---

*For this reason whenever we speak simply of binary or ternary forms we will always mean binary or ternary forms of the *second degree*.

where we have used in a special instance the matrix units $e_{hj}^{p\times q}$ which generally are defined as the matrices

$$e_{hj}^{p\times q} = (\delta_{ih}\delta_{kj}) \tag{2}$$

of p rows and q columns over a unital ring.

Thus forms and symmetric integral quadratic matrices may be used interchangeably. Note that the forms of n variables form a free module of rank $\binom{n+1}{2}$ under addition with the null form of n variables as natural element.

REPRESENTATION BY A FORM

A form (1a) is said to *represent* (or to *imply* or to *contain*) the form

$$g(y_1,\dots,y_m) = \sum_{i,k=1} b_{ik}y_iy_k \tag{3a}$$

if there is a linear substitution

$$x_k = \sum_{i=1}^{n} y_i s_{ik} \qquad (i = 1, 2, \dots, m) \tag{3b}$$

with rational integral coefficients s_{ik} such that there holds the polynomial identity

$$g(y_1,y_2,\dots,y_m) = f(x_1,x_2,\dots,x_n) \tag{3c}$$

This condition is tantamount to the matrix equation

$$A_g = SA_fS^T \tag{3d}$$

for the matrix

$$S = (s_{ik}) \in \mathbb{Z}^{m\ n} \tag{3e}$$

The presentation by means of (3b) is said to be *proper* if the greatest common divisor of the s_{ik}'s is 1, otherwise, it is said to be *improper*.

For example, the choice $S = I_n$ amounts to a proper presentation of f by f, the choice $S = kI_n$ $(1 < k \in \mathbb{Z})$ amounts to an improper presentation of k^2f by f, the choice $S = 0^{m\times n}$ amounts to the trivial presentation of the null form in m variables by f which is improper, the choice of $S \in \mathbb{Z}^{1\times n}$, $S \neq 0$, amounts to a nontrivial presentation of the unary form with matrix SA_fS^T by f which is tied up in a one-to-one fashion with the nontrivial presentation of the rational integer $f(s_{11}, s_{12}, \ldots, s_{1m})$ by f.

LEMMA 1. If f contains g and g contains h, then f contains h.

Proof. There are integral matrices R, S such that

$$A_g = RA_fR^T, \qquad A_h = SA_gS^T$$

hence

$$A_h = SRA_f(SR)^T$$

Thus the containment relation of the form g in the form f, symbolically denoted by setting

$$A_f \geq A_g \tag{4a}$$

makes the set of all symmetric integral matrices a partially ordered set (poset). The equivalence relation of this poset leads to the concept of (*arithmetical*) *equivalence* of forms

$$\begin{aligned} f \underset{\mathbb{Z}}{\sim} g \Leftrightarrow A_f \geq A_g \quad &\text{and} \quad A_g \geq A_f \\ \Leftrightarrow A_f \underset{\mathbb{Z}}{\sim} A_g& \end{aligned} \tag{4b}$$

with the usual properties of reflexivity, symmetry and transitivity so that the equivalence classes form a proper poset according to the partial ordering relation

$$(f) \geq (g) \Leftrightarrow A_f \geq A_g \tag{4c}$$

Here we denote by (f) the set of all forms equivalent to f.

Let us quote Gauss once again ([1], Sec. 278):

> Now to complete the theory of ternary forms we require a solution of the following problem. I, To find all representations of a given number by a given ternary form. II, To find all representations of a given binary form by a given ternary form. III, To judge whether or not two given ternary forms of the same determinant are equivalent and, if they are, to find all transformations of one into the other. IV, To judge whether or not a given ternary form implies another given form of a greater determinant and, if it does, to assign all transformations of the first into the second.

Replacing "ternary" by "n-ary" and "binary" by "m-ary" when $1 < m < n$ the Gauss quotation establishes the basis for a sensible research program for forms.

DEGENERATE FORMS

A form is said to be *degenerate* if it is contained in a form of a smaller number of variables n (see Gauss [1], end of Sec. 267). For example, the null forms are degenerate (the null form in one variable may be said to be contained in the "empty form" (form of 0 variables)).

The *rank* of a form is defined as the rank of its matrix. The *determinant* of a form is defined as the determinant of its matrix.

The *discriminant* of a form is defined as its negative determinant.

LEMMA 2. An n-ary form is degenerate precisely if its rank is less than n. It is nondegenerate precisely if its determinant (discriminant) is nonzero.

Proof. Let f be an n-ary form.

The solution space of the linear homogeneous diophantine matrix equation

$$xA_f = 0^{1\times n} \qquad (x \in \mathbb{Z}^{1\times n}) \tag{5a}$$

is a module with a solution basis $x_1, \ldots, x_{n-r}$ over $\mathbb{Z}$ where r is the rank of A_f. It is a purely embedded submodule of $\mathbb{Z}^{1\times n}$ such that its $\mathbb{Z}$-basis can be extended to a $\mathbb{Z}$-basis $x_1, \ldots, x_n$ of $\mathbb{Z}^{1\times n}$ which is tantamount to the statement that the matrix X with rows $x_1, x_2, \ldots, x_n$ belongs to the unit group $GL(n,\mathbb{Z})$ of $\mathbb{Z}^{n\times n}$. Note that a matrix of degree n over $\mathbb{Z}$ has an integral inverse if and only if its determinant is equal to ±1. From this reason such matrices are said to be *unimodular*.

It follows that

$$XA_fX^T = \begin{pmatrix} 0^{(n-r)\times(n-r)} & 0^{(n-r)\times r} \\ 0^{r\times(n-r)} & A^* \end{pmatrix} \tag{5b}$$

where A* is the nonsingular symmetric matrix

$$A^* = SA_fS^T \tag{5c}$$

of degree r over $\mathbb{Z}$ with S the r × n matrix formed by the rows $x_{n-r+1}, \ldots, x_n$. Hence $A_f \geq A^*$. But since

$$XA_fX^T = (0^{(n-r)\times r}I_r)A^*(0^{(n-r)\times r}, I_r)^T \tag{5d}$$

$$A_f = X^{-1}(XA_fX^T)X^{-T} \tag{5e}$$

it follows that $A^* \geq A_f$ so that

$$A^* \underset{\mathbb{Z}}{\sim} A_f \tag{5f}$$

Summing up our results we obtain

THEOREM 1. Equivalent forms have equal rank.

Every equivalence class contains nondegenerate forms.

Two nondegenerate forms, say f and g, are equivalent precisely if there is a unimodular matrix S satisfying (3d).

Two equivalent nondegenerate forms have the same determinant.

Any solution S of (3d) is unimodular in case both f and g are nondegenerate and equivalent.

The determinant of a form of n variables contained in the nondegenerate form of n variables is obtained by multiplication of a square number into the determinant of f.

It follows from Theorem 1 that the discriminant of a nondegenerate form depends only on its equivalence class; we call it the *discriminant of the class*.

PROPER EQUIVALENCE

Two nondegenerate forms are said to be *properly* equivalent if there is a unimodular matrix of determinant 1 transforming one into the other. If there is none, but if they are arithmetically equivalent, then they are said to be *improperly* equivalent.

We note that the unimodular matrices of degree n and determinant 1 form a subgroup $SL(n,\mathbb{Z})$ of index 2 in the full unimodular group of degree n.

Hence the nondegenerate forms of an equivalence class are divided up either into one or into two *proper* equivalence classes.

A given nondegenerate form f determines only one proper equivalence class precisely if there is a unimodular matrix of determinant -1 transforming it into itself. In this case we speak of an *ambiguous* form, determining an *ambiguous* equivalence class. If every integral matrix transforming f into itself has determinant 1, then f is nonambiguous. It determines two proper equivalence classes, viz., the set of forms

$$\{g \mid A_g = XA_fX^T \text{ and } X \in SL(n,\mathbb{Z})\}$$

$$\{h \mid A_h = D_{1,n}XA_f(D_{1,n}X)^T \text{ and } X \in SL(n,\mathbb{Z})\}$$

where

$$D_{1,n} = \begin{pmatrix} -1 & & & & \\ & 1 & & & \\ & & \cdot & & \\ & & & \cdot & \\ & & & & \cdot \\ & & & & & 1 \end{pmatrix} \tag{5g}$$

is a certain unimodular matrix of degree n and determinant -1.

Since for odd degrees n also $-I_n$ is a unimodular matrix of degree n and determinant -1 and, moreover,

$$(-I_n)A(-I_n)^T = A \tag{5h}$$

for any matrix A of degree n, it follows that every equivalence class of odd rank is ambiguous (see Gauss [1], Sec. 268).

ALGEBRAIC BACKGROUND

Let us briefly look at the algebraic background of the theory of forms.

The n-ary form f is said to contain the m-ary form g *rationally* (*really, in complex terms*) if the matrix equation (3d) can be satisfied by a matrix S with rational (real, complex number) coefficients. We write

$$A_f \underset{2}{\geq} A_g \qquad (A_f \underset{\mathbb{R}}{>} A_g,\ A_f \underset{\mathbb{C}}{\geq} A_g) \tag{6a}$$

for these relationships which are increasingly weaker than the containment over $\mathbb{Z}$. Each of them establishes a partial ordering of the set of symmetric integral matrices with corresponding equivalence relations.

$$f \underset{2}{\sim} g \qquad (f \underset{\mathbb{R}}{\sim} g,\ f \underset{\mathbb{C}}{\sim} g) \tag{6b}$$

We know from linear algebra that in each case equivalent forms have the same rank and that any two forms of the same rank are equivalent over $\mathbb{C}$.

An, form f is 2-equivalent to a form with diagonal matrix.

Any form f of rank r is $\mathbb{R}$-equivalent to precisely one of the forms $\pm f_{\sigma,r}$ $(0 < \sigma \leq r,\ r \equiv \sigma \pmod 2)$ or to $f_{0,r}$ where

$$f_{\sigma,r}(x_1,\ldots,x_r) = \sum_{i=1}^{\sigma} x_i^2 + 2 \sum_{j=1}^{(r-\sigma)/2} x_{\sigma+2j-1}x_{\sigma+2j} \tag{6c}$$

(The corresponding matrix is shown on p. 84!) The number σ is said to be the *index of inertia* of f (and of (f)).

If it is equal to the rank, then f is said to be *semidefinite*.

A semidefinite form is characterized as a form that cannot have both positive and negative values. It is said to be *positive semidefinite* if it does not have negative values. It is said to be *negative semidefinite* if it does not have positive values. It is said to be *positive definite* if it is positive semidefinite and presents 0 only trivially. This is tantamount to the statement that the form is nondegenerate and positive semidefinite. A form f is said to be *negative definite* if -f is positive definite.

A form is said to be *indefinite* if it is not definite.

The following lemma of linear algebra, which will not be proven here (see Bodewig), can be used to recognize the rational or real equivalence class of a form.

LEMMA 3. For any rational (real) symmetric matrix $A = (a_{ik})$ of degree n and rank r there is a sequence of r nonzero principal minors μ_1, $\mu_1\mu_2$, ..., $\mu_1\mu_2\cdots\mu_r$ of 1, 2, ..., r, each of which is contained in the next. The quadratic form $f(x_1,\ldots,x_n) = \Sigma_{i,k=1}^{n} a_{ik}x_ix_k$ corresponding to the matrix is rationally (really) equivalent to the diagonal form $\Sigma_{i=1}^{r}\mu_i x_i^2$ of r variables determined by the r minors.

Example. The ternary form $-x_2^2 + 2x_1x_2$ is really equivalent to the diagonal form $-x_1^2 + x_2^2$. It is of rank 2 and of inertia index 0, hence it is really equivalent to $f_{0,2}$.

A binary form is degenerate, definite or indefinite, depending on the sign of its discriminant being 0, -1, 1. It is null precisely if the rank is 0. It is positive semidefinite and degenerate if its discriminant is 0 and one of the diagonal coefficients is positive.

A ternary nondegenerate form f with negative discriminant either is positive definite or f is really equivalent to $-f_{1,3}$ (Gauss [4], Sec. 271). The second alternative arises precisely if A_f has at least one negative 2-minor.

A FUNDAMENTAL STATEMENT

One of the fundamental statements of the theory is (Gauss, Sec. 276):

THEOREM 2. The number of equivalence classes of all forms of a given nonzero determinant and rank is finite.

The proof is achieved by a method of reduction first explained in Sec. 272 of Gauss, which was later on formulated in geometrical language for $r = \sigma = 3$ by Seeber and Dirichlet; it was generalized to all cases by Hermite and Minkowski.

The work of those four geometers prepared the formation of a new mathematical discipline, the *geometry of numbers*.

This exposition presents Gauss' method in the language of the geometry of numbers.

THE GEOMETRIC TREATMENT

The geometric treatment hinges on the interpretation of a quadratic form f as a continuous mapping of the linear space $\mathbb{R}^{1\times n}$ of dimension n over the real number field $\mathbb{R}$ into $\mathbb{R}$ satisfying the parallelogram condition

$$f(x + y) + f(x - y) = 2f(x) + 2f(y) \qquad (A_f = (a_{ik}) \in \mathbb{R}^{n\times n}) \tag{7}$$

which is tantamount to

$$f(x) = f(x_1,\ldots,x_n) = \sum_{i,k=1}^{n} a_{ik}x_i x_k = xA_f x^T$$

and the symmetry condition (1b). It is convenient to associate with f the symmetric bilinear form

$$B_f(x,y) = \sum_{i,k=1}^{n} a_{ik}x_i y_k \qquad \begin{array}{l} x = (x_1,\ldots,x_n) \\ y = (y_1,\ldots,y_n) \end{array}$$

mapping the Cartesian product $\mathbb{R}^{1\times n} \times \mathbb{R}^{1\times n}$ into $\mathbb{R}$ such that

$$B_f(u + v,w) = B_f(u,w) + B_f(v,w) \tag{8}$$

$$B_f(\lambda u,w) = \lambda B_f(u,w) \tag{9}$$

$$B_f(u,w) = B_f(w,u) \tag{10}$$

$(u,\ v,\ w \in \mathbb{R}^{1\times n},\ \lambda \in \mathbb{R})$

$$f(x) = B_f(x,x) \tag{11}$$

$$B_f(x,y) = \frac{1}{2}\,(f(x + y) - f(x) - f(y)) \tag{12}$$

Example. 1. The Cartesian dot product

$$B(x,y) = \sum_{i=1}^{n} x_i y_i = x \cdot y \tag{13}$$

corresponds to the choice of $(a_{ik}) = I_n$.

2. The null form

$$B(x,y) = 0 \tag{14}$$

corresponds to the choice of $(a_{ik}) = 0^{n\times n}$.

3. The bilinear form

$$B_{\sigma,n}(x,y) = \sum_{i=1}^{\sigma} x_i y_i + \sum_{j=1}^{(n-\sigma)/2} (x_{\sigma+2j-1}y_{\sigma+2j} + x_{\sigma+2j}y_{\sigma+2j-1}) \tag{15}$$

$(0 \le \sigma \le [n/2],\ \sigma \equiv n \pmod 2)$ corresponds to the quadratic form $f_{\sigma,n}$ over $\mathbb{R}$ with matrix

$$A_{\sigma,n} = \begin{array}{r} \sigma \left\{ \right. \\ \\ n-\sigma \left\{ \right. \end{array} \begin{pmatrix} 1 & & & & & \\ & \ddots & & & & \\ & & 1 & & & \\ & & & \begin{pmatrix} 0 & 1 \\ 1 & 0 \end{pmatrix} & & \\ & & & & \ddots & \\ & & & & & \begin{pmatrix} 0 & 1 \\ 1 & 0 \end{pmatrix} \end{pmatrix}$$

Thus we establish *quadratic spaces* of n dimensions over $\mathbb{R}$, i.e., the linear space $\mathbb{R}^{1\times n}$ together with a bilinear form B on $\mathbb{R}^{1\times n}$, as the affine geometric background of the arithmetic theory of n-ary forms.

In this light a nondegenerate n-ary form f of inertia index σ and determinant

$$d = d(f) = \det(a_{ik}) = \det A_f \tag{16}$$

determines a matrix S_f of $SL(n,\mathbb{R})$ for which

$$A_f = \pm|d^{1/n}|S_f A_{\sigma,n} S_f^T \tag{17a}$$

$$f(x) = \pm|d(f)^{1/n}|f_{\sigma,n}(xS_f) \tag{17b}$$

The arguments on the right of (17b) are the n-rows xS_f forming the discrete submodule $\mathbb{Z}^{1\times n}S_f$ with the n basis elements $e_{1i}^{1\times n}S_f$ $(1 \le i \le n)$ over $\mathbb{Z}$.

The discrete submodules of $\mathbb{R}^{1\times n}$ are called *vector lattices.*

Discreteness is defined relative to the natural topology. In other words, a subset L of $\mathbb{R}^{1\times n}$ is said to be "discrete" if for each element x of L there is a positive real number ε such that for any element y of L distinct from x we have the inequality

$$(y - x)(y - x)^T > \varepsilon^2$$

or, negatively stated, the ε-ball centered at x:

$$\{y | y \in \mathbb{R}^{1\times n} \text{ and } (y - x)(y - x)^T \le \varepsilon^2\}$$

contains no element of L other than x.

A trivial example of a vector lattice is formed by $0^{1\times n}$, the 0 element of $\mathbb{R}^{1\times n}$ only. It is called the zero-dimensional lattice, or the *zero-lattice.*

LEMMA 4. Any nonzero vector lattice L of $\mathbb{R}^{1\times n}$ has a lattice basis b_1, b_2, ..., b_r such that

$$L = \mathbb{Z}b_1 \dot{+} \mathbb{Z}b_2 \dot{+} \cdots \dot{+} \mathbb{Z}b_r \tag{18}$$

and the r rows are linearly independent over $\mathbb{R}$. The number r is equal to the $\mathbb{R}$-dimension of the linear $\mathbb{R}$-space generated by L, a lattice invariant. The module structure of L is the structure of the free abelian group of r generators where $1 \le r \le n$. Conversely, any r linearly independent elements b_1, ..., b_r of $\mathbb{R}^{1\times n}$ span the r-lattice L according to (18).

Proof of Lemma 4. There is a positive real number ε such that for any nonzero element x of L we have $|x| = |(xx^T)^{1/2}| > \varepsilon$.

Since L is not zero and L is embedded into $\mathbb{R}^{1\times n}$ it follows that the maximum number of $\mathbb{R}$-linearly independent elements of L is a natural number r not greater than n. There are r $\mathbb{R}$-linearly independent elements $a_1, a_2, \ldots, a_r$ of L. For any two distinct elements x, y of L there hold equations

$$x = \sum_{i=1}^{r} \xi_i a_i \qquad y = \sum_{i=1}^{r} \eta_i a_i$$

with real numbers $\xi_1, \ldots, \xi_r, \eta_1, \ldots, \eta_r$ such that

$$\mu = \max_{1\le i\le r} |\eta_i - \xi_i| > 0$$

The nonzero difference y - x belongs to L so that

$$0 < \varepsilon < |y - x| = \left| \sum_{i=1}^{r} (\eta_i - \xi_i) a_i \right|$$

$$\le \sum_{i=1}^{r} |\eta_i - \xi_i| |a_i| \le \mu \sum_{i=1}^{r} |a_i|$$

$$\mu > \varepsilon \left(\sum_{i=1}^{r} |a_i| \right)^{-1} > \frac{1}{k} \tag{19}$$

where k is the natural number $1 + [\varepsilon^{-1}\Sigma_{i=1}^{r}|a_i|]$.

In order to show the existence of a lattice basis $b_1, \ldots, b_r$ we must show that the torsion-free additive abelian group L contains the free abelian group L_0 generated by $a_1, \ldots, a_r$ as a subgroup of finite index.

Indeed, every element

$$x = \sum_{i=1}^{r} \xi_i a_i \qquad \xi_i \in \mathbb{R},\ 1 \le i \le r$$

of L is congruent to the element $x' = x - \Sigma_{i=1}^{r}[\xi_i]a_i = \Sigma_{i=1}^{r}(\xi_i - [\xi_i])a_i$, modulo L_0, and each x' belongs to precisely one of the k^r boxes

$$B_{j_1,j_2,\ldots,j_r} = \left\{ \sum_{i=1}^{r} \beta_i a_i \,\middle|\, j_i - 1 \le k_{\beta_i} < j_i,\ 1 \le i \le r \right\}$$

($\beta_i \in \mathbb{R}$, $0 < j_i \le k$, $1 \le i \le r$). But there cannot be more than one element of L_0 in any one of the k^r boxes because of the inequality (19). Hence the index of L over L_0 is finite and not greater than k^r.

Remarks. (a) It follows from the basis theorem for free abelian groups of finite rank that all lattice bases of L are derived from a given lattice basis $b_1, \ldots, b_r$ by means of unimodular transformations of the form

$$c_i = \sum_{k=1}^{r} w_{ik} b_k \qquad 1 = 1, 2, \ldots, r$$

where $(w_{ik}) \in GL(r,\mathbb{Z})$.

(b) The lattice $\mathbb{Z}^{1\times n} S_f$ is n-dimensional for any nondegenerate n-ary form f.

Even though vector lattices are *discrete* subsets of $\mathbb{R}^{1\times n}$ yet one may think of them as to be dispersed over the affine space range more loosely or more densely.

As a measure for the looseness of a vector lattice L with lattice basis $b_1, b_2, \ldots, b_r$, one uses customarily the *mesh of the lattice*, i.e., the positive number

$$\Delta(L) = |\det(b_i b_k^T)^{1/2}| \tag{20a}$$

The r × n matrices C, the rows of which form the basis of an r-dimensional sublattice L of L, are obtained in the form

$$C = \Omega B$$

where B is the r × n matrix with row vectors $b_1, \ldots, b_r$ and Ω is a nonsingular matrix of degree r over $\mathbb{Z}$. Hence

$$L : L_1 = |\det \Omega| \tag{20b}$$

$$\begin{aligned}\Delta(L_1) &= |\det(CC^T)^{1/2}| \\ &= |\det(\Omega BB^T \Omega^T)^{1/2}| \\ &= |(\det \Omega)\det(BB^T)(\det \Omega^T))^{1/2}| \\ &= |\det \Omega|\,|\det(BB^T)^{1/2}|\end{aligned}$$

$$\Delta(L_1) = (L : L_1)\Delta(L) \tag{20c}$$

In particular, the mesh is independent of the choice of the basis used in (20a).

The mesh happens to be 1 in case $b_1, \ldots, b_r$ are distinct unit rows.

In other words, the mesh is the Euclidean r-dimensional volume of any r-dimensional parallelotope of the n-dimensional Euclidean space with edge vectors $b_1, \ldots, b_r$.

The n-dimensional mesh of the n-dimensional lattice $\mathbf{Z}^{1\times n}A_f$ is equal to $|d(f)^{1/2}|$ where f is any nondegenerate n-ary form.

THE GEOMETRIC TREATMENT

We compare the well-ordered subsets S, S_1, S_2, ..., of $\mathbf{R}^{1\times n}$ according to lexicographic Euclidean length ordering. Denote by $\Omega(S)$ the ordinality of S and denote by $S(\omega)$ the ωth element of S if ω is a nonlimit ordinal number $\leq \Omega(S)$. The well ordered subset S_1 of $\mathbf{R}^{1\times n}$ is said to be *lower* than the well-ordered subset S_2 of $\mathbf{R}^{1\times n}$ if either

$$\Omega(S_1) < \Omega(S_2)$$

and

$$S_1(\omega)S_1^T(\omega) = S_2(\omega)S_2^T(\omega)$$

(ω a nonlimit ordinal number $\leq \Omega(S_1)$), or if

$$\Omega(S_1) \leq \Omega(S_2)$$

and for some nonlimit ordinal number ω_0

$$\omega_0 \leq \Omega(S_1)$$

$$S_1(\omega)S_1^T(\omega) = S_2(\omega)S_2^T(\omega) \qquad 1 \leq \omega < \omega_0$$

$$S_1(\omega_0)S_1^T(\omega_0) \leq S_2(\omega_0)S_2^T(\omega_0)$$

It follows that for any two well-ordered subsets S_1, S_2 of $\mathbf{R}^{1\times n}$ precisely one of the three following relations holds; either

S_1 is lower than S_2

($S_1 < S_2$, $S_2 > S_1$, S_2 is *higher* than S_1)

or

S_2 is lower than S_1

($S_2 < S_1$, $S_1 > S_2$, S_1 is higher than S_2)

or

S_2 is *Euclidean equivalent* to S_1 which means that

$$\Omega(S_1) = \Omega(S_2)$$

and that

$$S_1(\omega)S_1^T(\omega) = S_2(\omega)S_2^T(\omega)$$

for any nonlimit ordinal $\omega \le \Omega(S_1)$.

If the square lengths xx^T of the elements x of a union of well-ordered subsets of $R^{1\times n}$ form a discrete set (e.g., if all of them are well-ordered subsets of a vector lattice), then there is a lowest well-ordered set among them which is unique up to equivalence.

For example, among the ordered r-tuples $(a_1,a_2,\ldots,a_r)$ of r independent elements $a_1, a_2, \ldots, a_r$ of an r-vector lattice L of $\mathbf{R}^{1\times n}$ there is a lowest one, say $(a_{01},a_{02},\ldots,a_{0r})$, such that the length squares

$$m_1(L) = a_{01}a_{01}^T,\ m_2(L) = a_{02}a_{02}^T,\ \ldots,\ m_r(L) = a_{0r}a_{0r}^T \tag{21a}$$

are uniquely determined. They are characterized by the conditions

$$0 < m_1(L) \le m_2(L) \le \cdots \le m_r(L) \tag{21b}$$

$a_{01}, \ldots, a_{0r}$ are linearly independent elements of the r-lattice L (21c)

$m_i(L) \le xx^T$ for any element of L that is linearly independent of the a_{0j} $(1 \le j \le i,\ i = 1, 2, \ldots, r)$ (21d)

The r numbers $m_1(L), m_2(L), \ldots, m_r(L)$ are called the *higher* (Euclidean distance) *square minima* of L (see Minkowski and Cassels).

Another example is obtained from the observation that among the ordered r-tuples $(b_1,b_2,\ldots,b_r)$ of a $\mathbb{Z}$-basis $b_1, \ldots, b_r$ of L there is a lowest one, say $(b_{01},b_{02},\ldots,b_{0r})$, such that the length squares

$$\ell_1(L) = b_{01}b_{01}^T,\ \ell_2(L) = b_{02}b_{02}^T,\ \ldots,\ \ell_r(L) = b_{0r}b_{0r}^T \tag{22a}$$

are uniquely determined. They are characterized by the condition

$$0 < \ell_1(L) \le \ell_2(L) \le \cdots \le \ell_r(L) \tag{22b}$$

$b_{01}, \ldots, b_{0r}$ form a $\mathbb{Z}$-basis of L (22c)

$\ell_i(L) \le xx^T$ for any element x of L that serves jointly with $b_{01}, \ldots, b_{0,i-1}$ as a $\mathbb{Z}$-basis of the intersection of L with the linear subspace $\mathbf{R}x \dotplus \Sigma_{j=1}^{i-1} \mathbf{R}b_{0j}$ so that the r elements $b_{01}, \ldots, b_{0i-1}, x$ can be extended to a $\mathbb{Z}$-basis of L (22d)

Note that

$$\ell_1(L) = m_1(L) \qquad \ell_i(L) \ge m_i(L) \qquad 1 \le i \le r \tag{22e}$$

Definition (Minkowski): The basis $b_1, \ldots, b_r$ of the r-lattice L is said to be *reduced* if it is lowest among the bases of L.

Note that $b_i b_i^T = \ell_i(L)$.

There is at least one reduced basis of L. There are only finitely many reduced bases of L.

FINDING A REDUCED BASIS OF A LATTICE

To find a reduced basis of an r-lattice L with basis $b_1, b_2, \ldots, b_r$ is trivial if r = 1, since each basis (b_1 or $-b_1$) is already reduced.

In order to solve the assignment in general it is useful to derive from the given basis $b_1, \ldots, b_r$ of the $\mathbb{R}$-linear space $\mathbb{R}L$ generated by L another basis $b_1', b_2', \ldots, b_r'$ for which

$$b_k' = \sum_{i=1}^{r} \lambda_{ik} b_i \qquad 1 \le k \le r \tag{24a}$$

where

$$(\lambda_{ik}) \in SL(r,\mathbb{R}) \tag{24b}$$

$$\lambda_{ii} = 1 \tag{24c}$$

$$\lambda_{ik} = 0 \qquad \text{if } 1 \le k < i \le r \tag{24d}$$

and the *orthogonality conditions*

$$b_i' b_k'^T = 0 \qquad i \ne k;\ i, k = 1, 2, \ldots, r \tag{24e}$$

are satisfied. The unique solution of this problem is obtained by the Gram-Schmidt orthogonalization which is defined recursively by

$$b_1' = b_1$$

$$b_i' = b_i - \sum_{j=1}^{i-1} (b_i b_j^T / b_j' b_j'^T) b_j' \qquad 1 \le i \le r \tag{24f}$$

Note that

$$\det(b_i b_k^T \mid i,k = 1,2,\ldots,s) = \det(b_i' b_k'^T \mid i,k = 1,2,\ldots,s) = \prod_{i=1}^{s} b_i' b_i'^T \tag{24g}$$

for s = 1, 2, ..., r.

Using the orthogonalized basis b_1', ..., b_r' we explicitly solve the task of obtaining for any element x of $\mathbb{R}^{1\times n}$ its *orthogonal projection*, x*, on the linear space $\mathbb{R}$L, i.e., we solve the system of inhomogeneous linear equations

$$b_i \sum_{k=1}^{r} \xi_k b_k^T = b_i x^T \qquad 1 \le i \le r \tag{25a}$$

for the unknown real coefficients ξ_1, ..., ξ_r of

$$x^* = \sum_{k=1}^{r} \xi_k b_k \tag{25b}$$

Since the determinant of the quadratic system of linear equations is the positive number $\det(b_i b_k^T)$, it is clear that there is precisely one solution. Employing b_1', ..., b_r' we obtain it in the form

$$x^* = \sum_{k=1}^{r} (b_k' x^T / b_k' b_k'^T) b_k \tag{25c}$$

The difference x - x* is the $\mathbb{R}$L-*normal component* of x characterized by the conditions

$$b_i (x - x^*)^T = 0 \qquad 1 \le i \le r \tag{25d}$$

that are equivalent to (25a) or to

$$\mathbb{R}L(x - x^*)^T = 0 \tag{25e}$$

and by the condition

$$x^* \in \mathbb{R}L \tag{25f}$$

equivalent to (25b).

It is useful to calculate the square of the volume of the (r+1)-parallelotope with the vectors b_1, b_2, ..., b_r, x in the form

$$\det(b_i b_k^T)(x - x^*)(x - X^*)^T = (\Pi_{i=1}^r b_i' b_i'^T)(x - x^*)(x - x^*)^T \tag{25g}$$

Note also the Pythagorean equation

$$xx^T = x^*x^{*T} + (x - x^*)(x - x^*)^T \tag{25h}$$

Assume now that $b_1, b_2, \ldots, b_{r+1}$ are the basis vectors of an (r+1)-vector lattice L_1 of the $\mathbb{R}^{1\times n}$ which we want to reduce, assuming there is a reduction algorithm for r-lattices already ($r \geq 1$).

Using the induction hypothesis we may assume without loss of generality that the r vectors $b_1, \ldots, b_r$ already form a reduced basis of the r-sublattice L which they generate.

Consider the set of all vectors

$$y = \sum_{i=1}^{r+1} \eta_i b_i \qquad \{\eta_i \in \mathbb{Z},\ 1 \leq i < r+1,\ \eta_{r+1} \neq 0,\ \gcd(\eta_1, \ldots, \eta_{r+1}) = 1\} \tag{26a}$$

There are two possibilities:

i. For all y of the form (26a) and for all indices h satisfying $1 \leq h \leq r+1$, and

$$\gcd(\eta_h, \eta_{h+1}, \ldots, \eta_{r+1}) = 1 \tag{26b}$$

it follows that

$$yy^T \geq b_h b_h^T \tag{26c}$$

The inequalities (26c) together imply that $b_1, b_2, \ldots, b_{r+1}$ form a reduced basis of L_1.

ii. For some vector of the form (26a) there is an index h satisfying (26b) and

$$yy^T < b_h b_h^T \tag{26d}$$

violating one of the reduction conditions. In the latter case the (r+2-h)-row

$$R_1 = (\eta_h, \eta_{h+1}, \ldots, \eta_{r+1})$$

can be augmented to a unimodular matrix (η_{ik}) of degree $r + 2 - h$ with R_1 as first row because of the condition (26b). Correspondingly, we find the new $\mathbb{Z}$-basis consisting of b_i ($1 \leq i < h$), y, and of the $r + 1 - h$ elements

$$\sum_{k=1}^{r+2-h} \eta_{ik} b_{h+k-1} \qquad 2 \le i \le r + 2 - h$$

which is lower than the given basis $b_1, \ldots, b_{r+1}$. Because of the discreteness of L_1 the operation ii can be performed only a finite number of times, whereafter we shall discover a reduced basis *provided we can establish a method by which only finitely many of the infinitely many inequalities* (26c) *need to be tested.*

In order to meet this requirement Dirichlet developed the concept of *Dirichlet region* which was rediscovered by many scientists under the name of a *domain of action* ('Wirkungsbereich') and again by Voronoj [7], so that the Dirichlet regions of lattice points with respect to the lattice to which they belong also is called its *Voronoj cell.* We define the Dirichlet region of an element x of the $\mathbb{R}^{1\times n}$ with respect to the nonempty subset S of $\mathbb{R}^{1\times n}$ as the set

$$\mathcal{D}(x,S) = \{y \mid y \in \mathbb{R}^{1\times n} \text{ and } \forall\, s(s \in S \Rightarrow (y - x)(y - x)^T \le (y - s)(y - s)^T)\}$$

In other words, $\mathcal{D}(x,S)$ consists of all elements of $\mathbb{R}^{1\times n}$ with Euclidean distance from x not greater than the distance from any element of S. Since the elements y of $\mathbb{R}^{1\times n}$, which are not farther removed from x in Euclidean terms than from the element $s \neq x$ of $\mathbb{R}^{1\times n}$, form the closed half-space containing x that is bounded by the median hyperplane of x and s, it follows that $\mathcal{D}(x,S)$ is the intersection of the closed half-spaces containing x and bisecting x from the elements of S. Hence, $\mathcal{D}(x,S)$ is a closed convex subset of $\mathbb{R}^{1\times n}$ containing x, but no element of S other than x.

The concept of a Dirichlet region is invariant under rigid motion. In particular, it is translation invariant. Since for any lattice L the translations carrying one lattice element into another one form an abelian normal subgroup of its symmetry group (which is also called the *translation group* of the lattice), it follows that those translations permute the Dirichlet regions of the elements of L with respect to L regularly and transitively.

Let $b_1, \ldots, b_r$ be a basis of the lattice L. Since the bisecting half-spaces of 0 and the elements $\pm b_i$ $(1 \le i \le r)$ containing 0 already intersect with $\mathbb{R}L$ in a bounded set, it follows from the discreteness of L that the Dirichlet region of 0 with respect to L is a convex cell

(*Dirichlet cell*) with bounded intersection $\mathcal{D}(0,L) \cap \mathbb{R}L$. The Dirichlet cells $\mathcal{D}(x,L)$ for x of L are obtained as the lattice translates of $\mathcal{D}(0,L)$ by x of L.

The Dirichlet region $\mathcal{D}(x,L)$ for x of L is the orthogonal $(n - r)$-cylinder with the Dirichlet cell $\mathcal{D}(x,L) \cap \mathbb{R}L$ as its basis.

Any two Dirichlet regions

$$\mathcal{D}(x,L) \qquad \mathcal{D}(y,L) \qquad\qquad x,\ y \in L,\ x \neq y$$

have only boundary points in common. If $x \in L$ and if two distinct points y, z of $\mathcal{D}(x,L)$ differ by an element of L, then both of them are on the boundary of $\mathcal{D}(x,L)$.

The r-dimensional volume of the bounded convex cell $\mathcal{D}(0,L) \cap \mathbb{R}L$ is equal to the mesh of L. This is because $\mathcal{D}(0,L)$ intersects with $\mathbb{R}L$ in

$$\bigcup_{i=1}^{\sigma} (x_i + \Pi) \cap \mathcal{D}(0,L)$$

where

$$\Pi = \Pi(b_1,\ldots,b_r) = \{x \mid x \in \mathbb{R}^{1\times n} \text{ and } x = \sum_{i=1}^{r} \xi_i b_i \text{ and } 0 \geq \xi_i < 1\ (1 \leq i \leq r)\}$$

is the basic parallelotope of L in $\mathbb{R}L$ and $x_1, \ldots, x_\sigma$ are finitely many elements of L such that the union of the translates

$$-x_i + ((x_i + \Pi) \cap \mathcal{D}(0,L)) = \Pi \cap (-x_i + \mathcal{D}(0,L))$$

is Π and any two of them only have boundary points in common. Hence the sum of their r-volumina is $\Delta(L)$ on the one hand; on the other hand, after suitable translation of each piece, that sum adds up to the r-volume of $\mathcal{D}(0,L) \cap \mathbb{R}L$.

If we know already the vertices of the Dirichlet cell $\mathcal{D}(0,L) \cap \mathbb{R}L$ then it is not difficult to discover the lattice points x_i that are used to cut up the Dirichlet cell $\mathcal{D}(0,L) \cap \mathbb{R}L$ into σ convex subcells which upon translation by $-x_i$ unite without overlapping to the basic parallelotope. Each of them is characterized by finitely many linear inequalities obtained by $-x_i$-translation of some of the linear inequalities characterizing $\mathcal{D}(0,L) \cap \mathbb{R}L$ together with some of the 2r linear inequalities characterizing Π. Consequently, it is easy to test to which of the translated subcells a given element of Π belongs.

Given an element $x = \Sigma_{i=1}^{r} \lambda_i b_i$ of $\mathbb{R}L$ ($\lambda_i \in \mathbb{R}$, $1 \le i \le r$), we decide first to which of the translated subcells the element

$$\sum_{i=1}^{r} (\lambda_i - [\lambda_i])b_i$$

of Π belongs. Suppose it does belong to the translated subcell $-x_i + (x_i + \Pi) \cap \mathcal{D}(0,L)$. Then it follows that $x \in \mathcal{D}(x_i + \Sigma_{i=1}^{r} [\lambda_i]b_i, L)$, or equivalently, that

$$x - x_i - \sum_{i=1}^{r} [\lambda_i]b_i \in \mathcal{D}(0,L)$$

The bounded convex cell $\mathcal{D}(0,L) \cap \mathbb{R}L$ is obtainable in terms of finitely many inequalities of the type

$$xx^T \le (b - x)(b - x)^T \qquad b \in L \tag{27}$$

Which?

First, let $b = \pm b_i$ $(1 \le i \le r)$. The corresponding 2r linear inequalities define a bounded convex cell of $\mathbb{R}L$ containing $\mathcal{D}(0,L) \cap \mathbb{R}L$ which is the convex hull of finitely many vertices $v_1', v_2', \ldots, v_{\mu'}'$. Only those linear inequalities need to be joined with the 2r inequalities of above that correspond to an element $b = \Sigma_{i=1}^{r} \lambda_i' b_i$ $(\lambda_i' \in \mathbb{Z}, 1 \le i \le r)$ for which

$$bb^T \le 2 \max_{1 \le j \le \mu'} v_j' v_j'^T \tag{28}$$

Using the orthogonalized basis $b_1', \ldots, b_r'$ we transform (28) into a system of finitely many linear inequalities that can be recursively solved to yield a finite number of r-tuples of integers $\lambda_1', \ldots, \lambda_r'$.

Eventually, a finite irredundant set of linear inequalities defining $\mathcal{D}(0,L)$ and a finite vertex set $\{v_1, \ldots, v_\mu\}$ of $\mathcal{D}(0,L) \cap \mathbb{R}L$ will be obtained.

The application to the task of reducing a basis $b_1, \ldots, b_{r+1}$ of the (r+1)-lattice L_1 is plain. We had assumed without loss of generality that $b_1, \ldots, b_r$ is a reduced basis of the r-sublattice $L = \Sigma_{i=1}^{r} \mathbb{Z}b_i$.

The orthogonal projection of an element $x = \Sigma_{i=1}^{r+1} \lambda_i b_i$ $(\lambda_i \in \mathbb{Z}, 1 \le i \le r + 1)$ of L_1 on L is

$$x^* = \lambda_{r+1} b_{r+1}^* + \sum_{i=1}^{r} \lambda_i b_i$$

hence

$$x - x^* = \lambda_{r+1}(b_{r+1} - b^*_{r+1})$$

and due to the Pythagorean equation

$$xx^T \geq (x - x^*)(x - x^*)^T = \lambda^2_{r+1}(b_{r+1} - b^*_{r+1})(b_{r+1} - b^*_{r+1})^T$$

We need to test only those reduction conditions (26b) for which

$$0 \leq (x - x^*)(x - x^*)^T < b_{r+1}b_{r+1}{}^T$$

and $\lambda_{r+1} > 0$, i.e.,

$$1 \leq \lambda_{r+1} < |(b_{r+1}b_{r+1}{}^T)/((b_{r+1} - b^*_{r+1})(b_{r+1} - b^*_{r+1})^T))^{1/2}| \tag{29}$$

For each of the finitely many choices of λ_{r+1} we apply the method given above to find an element $y_{\lambda_{r+1}}$ of L for which $\mathcal{D}(y_{\lambda_{r+1}},L)$ contains $\lambda_{r+1}b^*_{r+1}$. Now we need to test only the finitely many reduction conditions (26b) derived from

$$y = \lambda_{r+1}b_{r+1} - y_{\lambda_{r+1}}$$

Thus a reduction algorithm is obtained.

Examples. r = 1: b_1 is reduced already. r = 2: (i) If $b_1b_1{}^T > b_2b_2{}^T > 0$, interchange b_1, b_2. (ii) If $0 < b_1b_1{}^T \leq b_2b_2{}^T$ and if $(b_1 \pm b_2)(b_1 \pm b_2)^T \geq b_2b_2{}^T$, i.e.,

$$|b_1b_2{}^T| \leq \frac{1}{2} b_1b_1{}^T \tag{30a}$$

then already b_1, b_2 is a reduced basis. Otherwise, let r be the nearest rational integer to $(b_1b_2{}^T)/(b_1b_1{}^T)$, replace b_1, b_2 by b_1, $b_2 - rb_1$ and start all over reducing.

Supposing

$$b_1b_2{}^T \geq 0 \tag{30b}$$

then the Minkowski reducing inequalities are given by

$$xx^T \leq (b - x)(b - x)^T \quad \text{with } b = \pm b_1, \pm b_2, \pm b_1 - b_2 \tag{30c}$$

defining the orthogonal cylinder on the basis of the *Dirichlet hexagon*

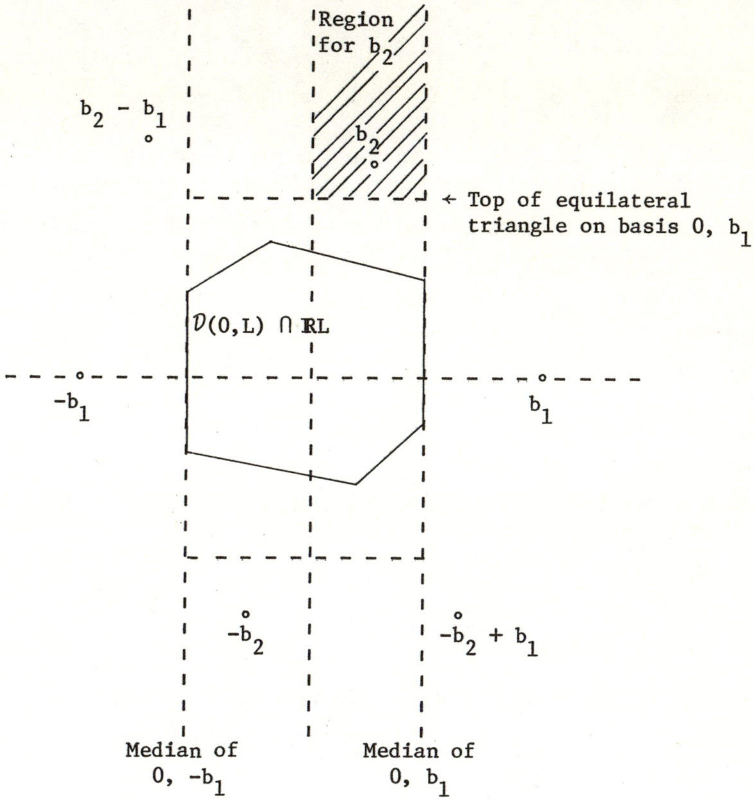

$\mathcal{D}(0,L) \cap \mathbb{R}L$ for b_1, b_2 (see figure): It degenerates to a rectangle in case b_1, b_2 are orthogonal. Note that (30a) implies that

$$\Delta(L) = \det(b_i b_k^T) = (b_1 b_1^T)(b_2 b_2^T) - (b_1 b_2^T)^2$$
$$\geq (b_1 b_1^T)(b_2 b_2^T) - \tfrac{1}{4}(b_1 b_1^T)^2$$

$$\Delta(L) \geq \tfrac{3}{4}(b_1 b_1^T)(b_2 b_2^T) \qquad r = 2 \tag{30d}$$

$$\Delta(L) = \tfrac{3}{4}(b_1 b_1^T)(b_2 b_2^T) \Leftrightarrow b_1 b_1^T = b_2 b_2^T = 2|b_1 b_2^T| \qquad r = 2 \tag{30e}$$

$r > 2$: (i) Reduce any subcouples b_i, b_k $(i < k)$ for which $|b_i b_k^T| > \frac{1}{2} b_i b_i^T$. (ii) If each of the $\binom{r}{2}$ subcouples b_i, b_k $(i < k)$ is reduced so that the inequalities

$$0 \le b_i b_1^T$$

$$b_i b_i^T \le b_{i-1} b_{i+1}^T \qquad 1 \le i < r \tag{30f}$$

$$|b_i b_k^T| \le \frac{1}{2} b_i b_i^T \qquad 1 \le i < k \le r$$

obtain, then we seek out among the other reducing inequalities all those which employ only the rational integers +1, -1, or 0:

$$\left(\sum_{i=1}^{r} \varepsilon_i b_i\right)\left(\sum_{i=1}^{r} \varepsilon_i b_i\right)^T \ge b_h b_h^T \tag{30g}$$

(where $\varepsilon_i = +1, 0$; $3 \le \Sigma_{i=1}^{r} |\varepsilon_i| \le r$; $h \le r$; $\varepsilon_h = 1$, $\varepsilon_i = 0$ if $h < i \le r$). How far do they get us?

$r = 3$: As we saw already the inequalities (30f), (30g) imply all inequalities

$$\left(\sum_{i=1}^{3} \eta_i b_i\right)\left(\sum_{i=1}^{3} \eta_i b_i\right)^T \ge b_3 b_3^T \tag{30h}$$

for rational integers η_1, η_2; η_3 satisfying $|\eta_3| = 1$.

We know from elementary geometry that the quotient of the radius of the circle circumscribing a triangle (without an obtuse angle) and the length of the second largest side is not greater than $\sqrt{2}/2$, the maximum being attained only for the isosceles right-angle triangle. It follows that

$$xx^T \le \frac{1}{2} b_2 b_2^T \qquad x \in \mathcal{D}(0,L),\ L = \mathbb{Z}b_1 + \mathbb{Z}b_2 \tag{30i}$$

with the sign of equality only if

$$b_1 b_1^T = b_2 b_2^T \qquad b_1 b_2^T = 0,\ x = \frac{1}{2}(\pm b_1 \pm b_2) \tag{30j}$$

Hence for a reduced basis b_1, b_2, b_3 of a 3-lattice L_1 with L as the sublattice

$$\begin{aligned} &b_3^* \in \mathcal{D}(0,L) \\ &b_3^* b_3^{*T} \le \frac{1}{2} b_2 b_2^T \\ &(b_3 - b_3^*)(b_3 - b_3^*)^T \ge b_3 b_3^T - \frac{1}{2} b_2 b_2^T \\ &(b_3 - b_3^*)(b_3 - b_3^*)^T \ge \frac{1}{2} b_2 b_2^T \end{aligned} \tag{30k}$$

where the sign of equality applies precisely if

$$0 < b_1b_1^T = b_2b_2^T = b_3b_3^T \qquad b_1b_2^T = 0$$
$$|b_1b_3^T| = |b_2b_3^T| = \frac{1}{2} b_2b_2^T \tag{30l}$$

Note that (30g) suffices to establish the remaining reducing inequalities for each of which $|\eta_3| \geq 2$. It follows that (30f), (30g) imply all other reducing inequalities. Note that because of (30f) we do not need to test (30g) unless

$$b_1b_2^T \times b_2b_3^T \times b_3b_1^T < 0 \tag{30m}$$

and in that case we need to test (30g) only if $\varepsilon_i = -\operatorname{sign} b_ib_3^T$ $(i = 1, 2)$.

$r = 4$: First we show the inequality

$$xx^T \leq \frac{3}{4} b_3b_3^T \qquad x \in \mathcal{D}(0,L_1) \cap \mathbb{R}L_1 \tag{30n}$$

where equality holds only if

$$0 < b_1b_1^T = b_2b_2^T = b_3b_3^T$$
$$b_1b_2^T = b_1b_3^T = b_2b_3^T = 0 \tag{30o}$$
$$|b_1x^T| = |b_2x^T| = |b_3x^T| = \frac{1}{2} b_1b_1^T$$

Indeed, setting $b_3 = b_3^* + n$, where b_3^* is the orthogonal projection of b_3 on $\mathbb{R}L$, we have $x = x^* + \lambda n$ for a certain real number λ. Always there hold the inequalities

$$x^*x^{*T} \leq \frac{1}{2} b_2b_2^T \qquad nn^T \leq b_3b_3^T$$

Hence if $|\lambda| \leq 1/2$, then (30n). Equality holds precisely if (30o) is satisfied.

If $|\lambda| > 1/2$, then the rational integer λ_0 nearest to λ is not zero. Since x belongs to $\dot{\mathcal{D}}(0,L_1)$ it follows that $xx^T \leq bb^T$, where $b = \eta_1b_1 + \eta_2b_2 - \lambda_0b_3 + x$ $(\eta_1, \eta_2 \in \mathbb{Z}, b^* \in \mathcal{D}(0,L_1))$, $xx^T \leq (\lambda - \lambda_0)^2nn^T + b^*b^{*T} \leq (1/2)^2b_3b_3^T + (1/2)b_3b_3^T = (3/4)b_3b_3^T$, which is (30n). Therefore, $|\lambda_0| = 1$. Hence equality in (30n) forces $|\lambda| = 1/2$.

It follows also that x of $\mathbb{R}L_1$ belongs to $\mathcal{D}(0,L_1)$ precisely if it satisfies the inequalities

$$xx^T \leq bb^T \qquad (b = x + \Sigma_{i=1}^{3} \eta_i b_i, \quad \eta_i = \pm 1, 0)$$

For any reduced basis b_1, b_2, b_3, b_4 of a 4-lattice L_2 we find that

$$b_4^* h_4^{*T} \leq \frac{3}{4} b_3 b_3^T$$

where b_4^* is the orthogonal projection of b_4 on the linear subspace $\mathbb{R}L_1$ generated by the 3-lattice with basis b_1, b_2, b_3. Hence $(b_4 - b_4^*) \times (b_4 - b_4^*)^T = b_4 b_4^T - b_4 b_4^T \geq b_4^* b_4^{*T} - (3/4) b_3 b_3^T$

$$(b_4 - b_4^*)(b_4 - b_4^*)^T \geq \frac{1}{4} b_4 b_4^T \tag{30p}$$

with the sign of equality only if both (30o) and

$$b_1 b_1^T = b_4 b_4^T = 2|b_1 b_4^T| = 2|b_2 b_4^T| = 2|b_3 b_4^T| \tag{30q}$$

The inequality (30p) implies the inequalities

$$b_4 b_4^t \leq bb^T$$

$$\text{for all } b = \sum_{i=1}^{4} \eta_i b_i, \ \eta_i \in \mathbb{Z} \ (1 \leq i \leq 4), \ |\eta_4| \geq 2 \tag{30r}$$

for every basis b_1, b_2, b_3, b_4 of L_2 with the property that b_1, b_2, b_3 form a reduced basis of the 3-lattice L_1 generated by them.

Using the remark made above we see that (30f), (30g) imply all other reducing inequalities.

$r = 5$: (30f), (30g) do not suffice as is shown by the 5-lattice with basis

$$e_i = e_{1i}^{1\times 5} \qquad 1 \leq i \leq 4$$

$$e_5 = \rho e_{15}^{1\times 5} + \frac{1}{2} \sum_{i=1}^{4} e_{1i}^{1\times 5} \qquad 0 < \rho < 1$$

Minkowski [5] gives a list of reduction inequalities employing the number 2 at most once each time and elsewhere ±1, 0.

The detailed inquiry into the cases $\gamma = 1, 2, 3, 4$ has shown that

$$m_i(L) = \ell_i(L) \qquad 1 \leq i \leq 4$$

for lattices of any dimension.

In practice a submodule of the $\mathbb{R}^{1\times n}$ is given as the row module R(B) generated by the n-rows

$$b_i = (\beta_{i1},\ldots,\beta_{in}) \qquad 1 \le i \le p$$

of a $p \times n$-matrix $B = (\beta_{ik})$ with real coefficients. The task is to decide whether R(B) is a sublattice of $\mathbb{R}^{1\times n}$ and in case of an affirmative answer to produce a *strongly reduced basis* $a_1, a_2, \ldots, a_{r(B)}$ of R(B). If the coefficients of B are rational, then R(B) is a lattice, otherwise the coefficients of B need to be given in a way that makes the decision possible. This reservation has to be made in regard to the algorithm which will follow.

The n-rows of $\mathbb{R}^{1\times n}$ are lexicographically ordered according to the total ordering relation stating that $(\xi_1,\ldots,\xi_n) \underset{\lambda}{<} (\eta_1,\ldots,\eta_n)$ if there is an index j for which $1 \le j \le n$, $\xi_j < \eta_j$, and $\xi_i = \eta_i$ if $1 \le i < j$.

A $\mathbb{Z}$-basis $a_1, a_2, \ldots, a_{r(B)}$ of R(B) is said to be strongly reduced if for any other $\mathbb{Z}$-basis $c_1, \ldots, c_{r(B)}$ of R(B) there is an index j for which $1 \le j \le n$ and either $c_j c_j^T > a_j a_j^T$ or $c_j \underset{\lambda}{>} a_j$, and $a_i = c_i$ if $1 \le i < j$. There is precisely one strongly reduced $\mathbb{Z}$-basis of R(B) if R(B) is a sublattice of $\mathbb{R}^{1\times n}$.

The aim is to produce a $p \times (n+p)$-matrix $M(B) = (P(B)B, P(B))$ with unimodular matrix B such that the first r(B) rows of P(B)B form the strongly reduced $\mathbb{Z}$-basis of R(B) and the remaining rows are zero. The exposition of the subsequent algorithm uses mathematical induction over p. We note that whenever a nondiscrete submodule of R(B) is found then also R(B) itself is nondiscrete. It is assumed tacitly that the submodules of R(B) obtained by the algorithm turn out to be discrete.

If p = 1 then either $B = 0^{1\times n}$, P(B) = (1), r(B) = 0, or there is an index j for which $1 \le j \le n$, $\beta_{1j} \ne 0$, $\beta_{1i} = 0$ if $1 \le i < j$. In the latter event we have $P(B) = (\operatorname{sign} \beta_{1j})$, r(B) = 1.

From now on let p > 1 and assume the algorithm is expounded already for rectangular matrices of less than p rows.

Start by setting $M(B) = (B, I_p)$. Denote by $\lambda(C)$ the last row of the matrix C, denote by $\underline{C}$ the matrix obtained by deletion of $\lambda(C)$, and denote by $\tau(C)$ the matrix obtained from C upon transposition of the last two rows.

1. If $\underline{P(B)B}$ is not strongly reduced then replace M(B) by $(P(\underline{P(B)B}) \oplus (1))M(B)$. After this replacement $\underline{P(B)B}$ is strongly reduced.

2. If $\lambda(P(B)B) = 0^{1\times n}$ then P(B)B is strongly reduced and $r(B) = r(P(B)B) = r(\underline{P(B)B})$.

3. If the hth row of $P(B)B$ is zero for some index h satisfying $1 \le h < p$, then transpose the pth and the hth row of $M(B)$ and apply 1, 2.

4. If $\underline{\tau(P(B)B)}$ is not strongly reduced then replace M by $\tau(M)$, apply 1 and then τ again. After the operations $\tau(P(T)B)$ will be strongly reduced.

5. If $\lambda\tau(P(B)B)\lambda\tau(P(B)B)^T \underset{\lambda}{>} \lambda(P(B)B)\lambda(P(B)B))^T$ then replace $M(B)$ by $\tau(M(B))$. Afterwards we have $\lambda\tau(P(B)B)\lambda\tau(P(B)B)^T \underset{\lambda}{\le} \lambda(P(B)B)\lambda(P(B)B)^T$.

6. If $\lambda(P(B)B) \underset{\lambda}{>} -\lambda(P(B)B)$ then multiply the last line of $M(B)$ by -1. Afterwards we have $\lambda(P(B)B) \underset{\lambda}{<} -\lambda(P(B)B)$.

Now assume that none of the 6 reductions 1-6 are applicable.

Let $a = \lambda\tau(P(B)B)$, $b = \lambda(P(B)B)$.

7. If $aa^T \le 2|ab^T|$ then there is the nonzero rational integer $\mu = -\operatorname{sign}(ab^T)[aa^T/2|ab^T|]$ such that $bb^T \ge (b + \mu a)(b + \mu a)^T$. If $bb^T > (b + \mu a)(b + \mu a)^T$ or if $bb^T = (b + \mu a)(b + \mu a)^T$ and $b \underset{\lambda}{>} b + \mu a$, $-(b + \mu a) \underset{\lambda}{>} (b + \mu a)$, then replace $M(B)$ by $(I_p + \mu e_{p,p-1})M(B)$. If $bb^T = (b + a) \times (b + a)^T$ and $b \underset{\lambda}{>} -(b + \mu a)$, $(b + \mu a) \underset{\lambda}{>} -(b + \mu a)$, then replace $M(B)$ by $(I_p + \mu e_{p,p-1})M(B)$, afterwards multiply the last row by -1. After this reduction 7 will not be applicable again because the submatrix formed by the two last rows of $P(B)B$ is strongly reduced. Thus $p = 2$ is settled.

Let $p > 2$. Assume 1-7 are not applicable.

By application of Gram-Schmidt's orthogonalization we construct vectors a^*, b^* of $\mathbb{R}^{1\times n}$ such that a^*, b^* are orthogonal to the first p-2 rows of $P(B)B$ and that the n-rows $a - a^*$, $b - b^*$ belong to the $\mathbb{R}$-linear space M spanned by the first p-2 rows of $P(B)B$. By construction a, b are nonzero. If a^*, b^* are linearly dependent over $\mathbb{R}$, say $b^* = \beta a^*$ and if β is irrational, then $R(B)$ is nondiscrete.

If, however, $b^* = \beta a^*$ and $\beta = \beta_1/\beta_2$ with nonzero mutually prime rational integers β_1, β_2, then we determine rational integers γ_1, γ_2 for which $\beta_1\gamma_2 - \beta_2\gamma_1 = 1$ and set $x = -\gamma_1 a + \gamma_2 b$. Then both $\beta_2 x - b$, $\beta_1 x - b$ belong to M. Applying the unimodular matrix

$$I_{p-3} \oplus \begin{pmatrix} \beta_1 & -\beta_1\gamma_1 & -\beta_1\gamma_2 \\ \beta_2 & -\beta_2\gamma_1 & -\beta_2\gamma_2 \\ 0 & -\gamma_1 & -\gamma_2 \end{pmatrix}^{-1}$$

to $M(B)$ and replacing $M(B)$ by the result we find 1, 2 to be applicable.

Now let a, b be linearly independent over $\mathbb{R}$.

In the sequel we must keep in mind the M-normal of any linear combination $\Sigma_{i=1}^{p}\xi_i c_i$ of the p rows $c_1, \ldots, c_p$ of P(B)B with real coefficients $\xi_1, \xi_2, \ldots, \xi_n$ is equal to $\xi_{p-1}a^* + \xi_p b^*$ and consequently the square length is at least as large as the square length of $\xi_{p-1}a^* + \xi_p b^*$.

8. In customary fashion we determine the finite set

$$S = \{\alpha a + \beta b \mid \alpha \in \mathbb{Z},\ \beta \in \mathbb{Z} \text{ and } (\alpha a^* + \beta b^*)(\alpha a^* + \beta b^*)^T \le aa^T \text{ and } \gcd(\alpha, \beta) = 1\}$$

For any member s of S distinct from ±a we form the (p-1) × n-matrix X(s) with rows $c_1, \ldots, c_{p-2}, s$. If there is an index j such that $1 \le j < p$, the ith row of P(X(s))X(s) is equal to c_i if $1 \le i < j$ and for the jth row c_j' of P(X(s))(s) we have $c_j'c_j'^T < c_jc_j^T$ or $c_j \underset{\lambda}{<} c_j$, then we determine two rational integers γ, δ such that $\alpha\delta - \beta\gamma = 1$ and apply the unimodular matrix

$$I_{p-2} \oplus \begin{pmatrix} \alpha & \beta \\ \gamma & \delta \end{pmatrix}$$

to M(B). Replacing M(B) by the resulting application we obtain a reduction by going through 1-7. Having done this a finite number of times, no further reduction 1-8 can be made. Thus for any element s of S there hold the inequalities

$$\left(\left(\sum_{i=1}^{p-2} \xi_i c_i\right) + s\right)\left(\left(\sum_{i=1}^{p-2} \xi_i c_i\right) + s\right)^T \ge aa^T \qquad (*)$$

$$(\xi_i \in \mathbb{Z},\ 1 \le i \le p - 2)$$

and in the case the sign of equality holds in (*) then we have

$$\left(\sum_{i=1}^{p-2} \xi_i cj\right) + s \underset{\lambda}{>} a$$

or else s = a.

By the same method we determine the set

$$S_1 = \{\pm(\alpha a + b) \mid \alpha \in \mathbb{Z} \text{ and } (\alpha a^* + b)(\alpha a^* + b)^T \le bb^T\}$$

For any number s of S distinct from ±b we form the (p-1) × n-matrix X(s) with rows $c_1, \ldots, c_{p-2}, s$.

If there is an index j such that $1 \le j < p$, the ith row of $P(X(s))x(s)$ is equal to c_i, if $1 \le i < j$ and for the jth row c_j of $P(X(s))X(s)$ we have $c_j' c_j'^T < c_j c_j^T$ or $c_j' \underset{\lambda}{<} c_j$ then we apply the unimodular matrix $I_p + \alpha e_{p,p-1}$ to M(B). Replacing M(B) by the resulting application we obtain a reduction by going through 1-8. Having done this a finite number of times, we find that 1-8 can no longer be applied. At this point P(B)B is strongly reduced.

Example.

$$B = \begin{array}{rrrr} -3 & 0 & -3 & 0 \\ -4 & -1 & 4 & 5 \\ -2 & -3 & -1 & 1 \\ 9 & 4 & 0 & -6 \end{array} \qquad M(B) = \left.\begin{array}{rrrr} -3 & 0 & -3 & 0 \\ -4 & -1 & 4 & 5 \\ -2 & -3 & -1 & 1 \\ 9 & 4 & 0 & -6 \end{array}\right| \begin{array}{rrrr} 1 & & & \\ & 1 & & \\ & & 1 & \\ & & & 1 \end{array}$$

$$\begin{array}{r} \\ \\ \rightarrow \\ \\ \end{array} \left.\begin{array}{rrrr} -3 & 0 & -3 & 0 \\ -2 & 2 & 5 & 4 \\ -2 & -3 & -1 & 1 \\ -3 & 4 & -6 & -6 \end{array}\right| \begin{array}{rrrr} 1 & & & \\ & 1 & -1 & \\ & & 1 & \\ 5 & & & 1 \end{array} \qquad \begin{array}{r} \\ \\ \rightarrow \\ \\ \end{array} \left.\begin{array}{rrrr} -1 & 3 & -2 & -1 \\ -2 & 2 & 5 & 4 \\ -2 & -3 & -1 & 1 \\ 1 & 6 & -1 & -2 \end{array}\right| \begin{array}{rrrr} 1 & 0 & -1 & 0 \\ 0 & 1 & -1 & 0 \\ 0 & 0 & 1 & 0 \\ 5 & 1 & -1 & 1 \end{array}$$

$$\begin{array}{r} \\ \\ \rightarrow \\ \\ \end{array} \left.\begin{array}{rrrr} -1 & 3 & -2 & -1 \\ -2 & 2 & 5 & 4 \\ -2 & -3 & -1 & 1 \\ 2 & 3 & 1 & -1 \end{array}\right| \begin{array}{rrrr} 1 & 0 & -1 & 0 \\ 0 & 1 & -1 & 0 \\ 0 & 0 & 1 & 0 \\ 4 & 1 & 0 & 1 \end{array} \quad \begin{array}{r} \\ \rightarrow \\ \\ \\ \end{array} \left.\begin{array}{rrrr} -1 & 3 & -2 & -1 \\ -2 & 2 & 5 & 4 \\ -2 & -3 & -1 & 1 \\ 0 & 0 & 0 & 0 \end{array}\right| \begin{array}{rrrr} 1 & 0 & -1 & 0 \\ 0 & 1 & -1 & 0 \\ 0 & 0 & 1 & 0 \\ 5 & 1 & 0 & 1 \end{array}$$

$$\begin{array}{r} \\ \\ \rightarrow \\ \\ \end{array} \left.\begin{array}{rrrr} 1 & -3 & 2 & 1 \\ 2 & 3 & 1 & -1 \\ 2 & -2 & -5 & -4 \\ 0 & 0 & 0 & 0 \end{array}\right| \begin{array}{rrrr} -1 & 0 & 1 & 0 \\ 0 & 0 & -1 & 0 \\ 0 & -1 & 1 & 0 \\ 5 & 1 & 0 & 1 \end{array} \qquad r(B) = 3$$

As the example shows, it pays to not be dogmatic. In the light of the previous exposition probably one should first apply all possible 2, 3, 4 reductions. Then perhaps one should divide up the matrix into two parts of $[p/2]$, $p - [p(2)]$ rows, respectively, and reduce the submatrices first, continuing afterwards with other subdivisions of the same kind until the ground is well enough prepared to make the ultimate test for strong reduction. Experience has shown that almost all matrices for which no 2-reductions can be applied are strongly reduced. Probability theory bears out this experience (see Donaldson 3).

PROOF OF THEOREM 2

The proof of Theorem 2 hinges on the proof of an inequality of the form

$$\prod_{i=1}^{r} m_i(L) \leq \varkappa_r \Delta(L)^2 \tag{32}$$

with a certain positive constant $\varkappa_r$ depending only on r, together with an inequality of the form

$$L : L_1 \leq \varkappa_r' \tag{33a}$$

for the index of any sublattice L_1 generated by r linearly independent elements $a_1, \ldots, a_r$ of the r-lattice L satisfying

$$a_i a_i^T = m_i(L) \qquad 1 \leq i \leq r \tag{33b}$$

where $\varkappa_r'$ is another positive constant depending on r only.

Both statements together imply that the absolute values of the coefficients of a reduced basis of an n-lattice of given mesh Δ are bounded by $\varkappa_n''|\Delta^{1/n}|$ with positive constant $\varkappa_n''$ depending on n only.

Applying that statement to a lattice L corresponding to an n-ary form of nonzero determinant d, it follows that the mesh of L is equal to $|d^{1/2}|$, so that f is equivalent to an n-ary form with matrix coefficients bounded by $n\,\varkappa_n''^2|d^{1/2}|$. Since there are only finitely many bounded integral matrices of degree n, it follows that the number of equivalence classes for given nonzero determinant and rank is finite.

Turning to (32), (33a) let us choose r linearly independent elements $a_1, \ldots, a_r$ of the r-lattice L satisfying (33b). By Gram-Schmidt we orthogonalize them to $a_1', a_2', \ldots, a_r'$. According to the definition of the higher minima, an element

$$a = \sum_{i=1}^{r} \alpha_i a_i' \qquad \alpha_i \in \mathbb{R},\ 1 \leq i \leq r$$

satisfying the inequalities

$$0 < \sum_{i=1}^{r} \alpha_i^2 \frac{a_i' a_i'^T}{a_i a_i^T} < 1 \tag{33c}$$

never can be a lattice vector. It follows that the linear transformation σ carrying a_i' on $|(a_i a_i^T)^{1/2}|a_i'$ maps L onto an r-lattice $L\sigma$ for which

$m_i(L\sigma) \ge 1$ $(1 \le i \le r)$. Note that the determinant of σ is equal to $\Pi_{i=1}^{r}|(a_i a_i^T)^{1/2}|$ and that $\mathcal{D}(0,L\sigma)$ contains the 1/2 ball centered at 0. Hence $\Delta(L\sigma) \ge \gamma_r 2^{-r}$ where γ_r is the r-volume of the unit ball in r dimensions. Thus $\varkappa_r = (2^r/\gamma_r)^2$ will do in order to get (32) established. The optimal values of $\varkappa_r$ for r = 1, 2, 3, 4 are given by the following table

r	$\varkappa_r^{-1}$	Critical lattice
1	1	e_{11}
2	3/4	e_{11}, $(e_{11}^{1\times 2} + \sqrt{3}e_{12}^{1\times 2})/2$
3	1/2	$e_{11}^{1\times 3}$, $e_{12}^{1\times 3}$, $\frac{1}{2}(e_{11}^{1\times 3} + e_{12}^{1\times 3} + \sqrt{2}e_{13}^{1\times 3})$
4	1/4	$e_{11}^{1\times 4}$, $e_{12}^{1\times 4}$, $e_{13}^{1\times 4}$, $\frac{1}{2}\Sigma_{i=1}^{4} e_{1i}^{1\times 4}$

which, however, does not give the only reduced basis. Apart from trivial modification via monomial basis transformation, one has for r = 3 also the "hexagonal lattice basis" $e_{11}^{1\times 3}$, $e_{11}^{1\times 3}/2 + \sqrt{3}e_{12}^{1\times 3}/2$,

$$\frac{1}{2}\, e_{11}^{1\times 3} + \sqrt{3}e_{12}^{1\times 3}/6 + \sqrt{2}e_{13}^{1\times 3}/\sqrt{3}$$

of the "critical lattice." The critical lattices given above correspond to the optimal positive definite reduced forms with matrices

$$(1),\ \begin{pmatrix} 2 & 1 \\ 1 & 2 \end{pmatrix},\ \begin{pmatrix} 2 & 0 & 1 \\ 0 & 2 & 1 \\ 1 & 1 & 2 \end{pmatrix},\ \text{or}\ \begin{pmatrix} 2 & 1 & 1 \\ 1 & 2 & 1 \\ 1 & 1 & 2 \end{pmatrix},\ \begin{pmatrix} 2 & 0 & 0 & 1 \\ 0 & 2 & 0 & 1 \\ 0 & 0 & 2 & 1 \\ 1 & 1 & 1 & 2 \end{pmatrix}$$

The optimality of $\varkappa_3 = 2$ is tantamount to the equality

$$8(s_2^2 - R^2)T^2 \ge s_1^2 s_2^4 \tag{33d}$$

of elementary geometry relating to a triangle with sides s_1, s_2, s_3 $(s_1 \le s_2 \le s_3)$, area T and radius R of the least circumscribed circle.

Examples for Theorem 2. For d = 1, σ = 3 the class number is 1 with I_3 as representative, for d = 1, σ = 1 the class number is 1 with $e_{11}^{3\times 3} - e_{22}^{3\times 3} - e_{33}^{3\times 3}$ as representative (see Gauss [4], Sec. 277).

THE AUTOMORPH

The automorph of an n-ary form f is defined as the semigroup

$$\mathrm{Aut}(f) = \mathrm{Aut}(f/\mathbb{Z}) = \{X \mid X \in \mathbb{Z}^{n\times n} \text{ and } XA_fX^T = A_f\} \tag{34}$$

of all matrices of degree n over $\mathbb{Z}$ transforming f on f. In other words, it consists of the matrices occurring in the presentations of f by f.

If f is nondegenerate then its automorph is a subgroup of the unimodular group $GL(n,\mathbb{Z})$, and conversely. Note that $\pm I_n$ always belongs to the automorph of f. The automorphs of two equivalent nondegenerate forms are conjugate under the unimodular group. The automorph remains unchanged upon multiplication of f by a nonzero rational integer.

For example, in case f is the square sum form its automorph is the *hyperoctahedral group* consisting of the $n!2^n$ monomial matrices of degree n over $\mathbb{Z}$.

Let the nondegenerate form f present the nondegenerate form g according to (3d), then we derive the presentations

$$XA_gX^T = XSY^{-1}(YA_fY^T)(XSY^{-1})^T \qquad X \in GL(m,\mathbb{Z}),\ Y \in GL(n,\mathbb{Z}) \tag{35a}$$

of the n-ary nondegenerate forms equivalent to g by the n-ary nondegenerate forms equivalent to f. In all, they form the *class of presentations equivalent to* (3d). It includes the subset of all those presentations (3a) for which $X \in \mathrm{Aut}(g)$, $Y \in \mathrm{Aut}(f)$ according to

$$A_g = XSY^{-1}A_f(XSY^{-1})^T \tag{35b}$$

In case one confines oneself to matrices X of $SL(m,\mathbb{Z})$, Y of $SL(n,\mathbb{Z})$ one speaks of *proper equivalence classes* of presentations. This distinction plays a role only if at least one of the numbers m, n is even. An equivalence class of presentations splits into 1, 2, or 4 proper equivalence classes according to the following table.

There is a remarkable duality which is exploited by Gauss in order to link the equivalence class of the proper presentations of a nonzero number D by a nondegenerate ternary form and the presentation of a binary form of determinant D by its adjoint.

The *adjugate* of a quadratic matrix $A = (a_{ik})$ of degree $n > 1$ over a commutative ring R is defined as the transpose of the algebraic complements $A' = (A_{ki})$ of the elements a_{ik} of A, where A_{ik} is the product of

Number of proper equivalence classes in a given equivalence class of presentations of the m-ary form g by the n-ary form f	Descriptive comment
1	$\mathrm{Aut}(g/\mathbb{Z}) \not\subseteq SL(m,\mathbb{Z})$, $\mathrm{Aut}(f/\mathbb{Z}) \not\subseteq SL(n,\mathbb{Z})$
2	$\mathrm{Aut}(g/\mathbb{Z}) \subseteq SL(m,\mathbb{Z})$, $\mathrm{Aut}(f/\mathbb{Z}) \not\subseteq SL(n,\mathbb{Z})$ or $\mathrm{Aut}(g/\mathbb{Z}) \not\subseteq SL(m,\mathbb{Z})$, $\mathrm{Aut}(f/\mathbb{Z}) \subseteq SL(n,\mathbb{Z})$
4	$\mathrm{Aut}(g/\mathbb{Z}) \subseteq SL(m,\mathbb{Z})$, $\mathrm{Aut}(f/\mathbb{Z}) \subseteq SL(n,\mathbb{Z})$

$(-1)^{i+k}$ and the (n-1)-minor, obtained by deleting the ith row and the kth column of A.

As is well known, this matrix operation satisfies the rules

$$AA' = (\delta_{ik} \det A) = (\det A) I_n = A'A \tag{36a}$$

$$(AB)' = B'A' \tag{36b}$$

characterizing the adjugate in case R is unital. Also there hold the rules

$$\det A' = (\det A)^{n-1} \tag{36c}$$

$$(A')' = (\det A)^{n-2} A \tag{36d}$$

$$(A')^T = (A^T)' \tag{36e}$$

The adjugate of a symmetric matrix is symmetric. The adjugate of an integral symmetric matrix is integral symmetric. The *adjoint* of the form f with matrix A_f is defined as the form ad(f) with matrix A_f' of the same degree as the degree of A_f.

The adjoint mapping preserves the partial ordering relation of the forms.

Indeed, by taking adjugates on both sides we infer from (3d) that

$$A_g' = S'^T A_f' S' = S'^T A_f' (S'^T)^T$$

Thus we may speak of the *adjoint equivalence class* ad((f)) = (ad(f)) of the equivalence class of the form f.

LEMMA 5. For n > 2 there is the one-to-one correspondence between the equivalence classes of proper presentations

$$W A_f' W^T = d \qquad W \in \mathbb{Z}^{1\times n} \tag{37a}$$

of the nonzero rational integer d by the adjoint of the n-ary nondegenerate form f and the proper equivalence classes of presentations

$$SA_fS^T = A_g \qquad S \in \mathbb{Z}^{(n-1)\times n} \tag{37b}$$

of (n-1)-ary forms g of determinant d by f such that

$$\binom{S}{0^{1\times n}}' = 0^{n\times(n-1)}W^T \tag{37c}$$

Proof. Indeed, starting with (37b) we set

$$\bar{S} = \begin{pmatrix} S \\ 0^{1\times n} \end{pmatrix} \tag{37d}$$

and find that

$$\bar{S}' = 0^{n\times(n-1)}W^T \tag{37e}$$

with $W \in \mathbb{Z}^{1\times n}$; moreover (37b) implies

$$\bar{S}A_f\bar{S}^T = \begin{pmatrix} A_g & 0 \\ 0 & 0 \end{pmatrix}$$

which implies (37a) upon taking adjugates on both sides.

Upon transition to an equivalent presentation

$$XA_gX^T = XSY^{-1}(YA_fY^T)(XSY^{-1})^T \qquad X \in GL(n-1,\mathbb{Z}),\ Y \in GL(n,\mathbb{Z})$$

we obtain

$$\overline{XSY^{-1}} = \begin{pmatrix} XSY^{-1} \\ 0^{1\times n} \end{pmatrix} = \tilde{X}\bar{S}Y^{-1} \qquad \tilde{X} = \begin{pmatrix} X & 0 \\ 0 & 1 \end{pmatrix}$$

$$\overline{XSY^{-1}}' = (Y')^{-1}\bar{S}'X' = (0^{n\times(n-1)}Y'^{-1}W^T \det(X)$$

leading to the presentation

$$W_1(YA_gY^T)'W_1{}^T = d$$

$$(W_1 = (\det X)W(Y'^T)^{-1})$$

of d by the adjugate of YA_gY^T which happens to be equivalent to (37a).

Note that the greatest common divisor of the coefficients of W is 1 if and only if the greatest common divisor of the coefficients of S is 1. Conversely, if there is a proper presentation (37a) of d by the adjoint of f then we construct presentations (37b) as follows. There exists a matrix V_0 of $SL(n,\mathbb{Z})$ such that

$$WV_0 = (0,0,\ldots,0,d)$$

All solutions of the matrix equation

$$P' = (0^{n\times(n-1)}(WV_0)^T) = V_0^T(0^{n\times(n-1)}W^T)$$

are obtained in the form

$$P = \begin{pmatrix} P_{11} & 0^{(n-1)\times 1} \\ 0^{1\times(n-1)} & 0 \end{pmatrix} \qquad P_{11} \in \mathbb{Z}^{(n-1)\times(n-1)},\ \det P_{11} = d$$

giving rise to the solutions

$$\bar{S} = P(V_0^{-T})'$$

of (37e), leading back to (37d), (37c), (37b).

PRIMITIVE BINARY FORMS

Given a form f then the greatest common divisor of the coefficients of f is invariant under equivalence. If it is 1 then f is said to be *primitive*.

Let

$$f(x,y) = px^2 + 2qxy + ry^2 \tag{38a}$$

be a primitive binary form of discriminant.

$$D = q^2 - pr \neq 0 \tag{38b}$$

$$p,\ q,\ r \in \mathbb{Z} \qquad \gcd(p,q,r) = 1 \tag{38c}$$

If f represents zero nontrivially then there is also a proper zero presentation. Hence f is equivalent to a primitive binary form

$$f'(u,v) = 2q'uv + r'v^2$$

It has the same discriminant as f, so that $D = (1')^2$. Conversely, if D is a square number, then indeed f represents zero nontrivially. Namely, if $p \neq 0$, then we have

$$pf(x,y) = (px + qy)^2 - Dy^2 \tag{38d}$$

and

$$D = k^2 \tag{38e}$$

therefore,

$$pf(x,y) = (px + (q + k)y)(px + (q - k)y) \tag{38f}$$

so that

$$f(-q-k,p) = f(-q+k,p) = 0$$

Now let us assume that D is a nonsquare number. This is tantamount to the statement that f represents zero only trivially. From (38d) we infer that

$$pf(x,y) = (px + qy + \sqrt{D}y)(px + qy - \sqrt{D}y) = N(px + (q + \sqrt{D})y)$$

where N denotes the absolute norm defined on the quadratic number field

$$K = 2(\sqrt{D}) \tag{38g}$$

by setting

$$\begin{aligned} N(a + b\sqrt{D}) &= N_{K/2}(a + b\sqrt{D}) = (a + b\ D)(a + b\sqrt{D})' \\ &= (a + b\sqrt{D})(a - b\sqrt{D}) = a^2 - b^2D \qquad (a,\ b \in 2) \end{aligned} \tag{38h}$$

using the nontrivial automorphism

$$\alpha \to \alpha'$$

of K that maps $\sqrt{D}$ on $-\sqrt{D}$. In other words, pf is the *norm form* on the free submodule

$$M_f = \mathbf{Z}p + \mathbf{Z}(q + \sqrt{D}) \tag{38i}$$

of K of rank 2. Conversely, the norm function on any free submodule $M = Z\alpha \dot{+} Z\beta$ of K of rank 2 defines the rational binary quadratic form

$$g_{\alpha,\beta}(x,y) = N(\alpha x + \beta y) = p_1x^2 + 2q_1xy + r_1y^2 \tag{38j}$$

with

$$\begin{aligned} p_1 &= N(\alpha) \qquad & q_1 &= (\alpha\beta' + \alpha'\beta)/2 = \tfrac{1}{2}\,\mathrm{tr}(\alpha\beta') \\ r_1 &= N(\beta) \qquad & g_{\alpha,\beta}(x,y) &= d^{-1}f_{\alpha,\beta}(x,y) \end{aligned} \tag{38k}$$

$f_{\alpha,\beta}$ primitive, d uniquely determined by α, β as the positive least common multiple of the least denominators of p_1, q_1, r_1.

In this way we establish a connection between the primitive forms of a given nonsquare discriminant D and the norm forms on the *Dedekind submodules* (free submodules of maximal rank) of the quadratic number field $2(\sqrt{D})$.

Let us point out that $f = p^{-1}f_{p,q+\sqrt{D}}$. Also let us point out that for a given Dedekind submodule

$$M = \mathbb{Z}\alpha + \mathbb{Z}\beta$$

of $2(\sqrt{D})$ the transition to a new module basis

$$\alpha\sigma = \alpha s_{11} + \beta s_{12}$$

$$\beta\sigma = \alpha s_{21} + \beta s_{22}$$

$$(s_{ik}) \in GL(2,\mathbb{Z}) \qquad (s_{ik}) \in SL(2,\mathbb{Z})$$

gives rise to a primitive form $f_{\alpha\sigma,\beta\sigma}$ that is equivalent (properly equivalent) to $f_{\alpha,\beta}$. Furthermore, for any λ of K of positive norm the transition from $M = \mathbb{Z}\alpha \dotplus \mathbb{Z}\beta$ to the Dedekind module $M\lambda = \mathbb{Z}\alpha\lambda \dotplus \mathbb{Z}\beta\lambda$ gives rise to the same primitive form $f_{\alpha,\beta}$ as we had seen before.

Finally, the transition from M to the algebraic conjugate Dedekind module

$$M' = \mathbb{Z}\alpha' \dotplus \mathbb{Z}\beta'$$

gives rise to the primitive form $f_{\alpha',\beta'}$ that is improperly equivalent to $f_{\alpha,\beta}$. This is because the determinant of the linear transformation $\xi \to \xi'$ $(\xi \in K)$ of K over 2 is -1 since the basis element 1 is kept fixed, but the basis element $\sqrt{D}$ is carried to its negative.

Conversely, if for two Dedekind submodules

$$M_j = \mathbb{Z}\alpha_j \dotplus \mathbb{Z}\beta_j \qquad j = 1, 2$$

of K the primitive forms f_{α_j,β_j} are equal, then either

$$\alpha_2\alpha_1^{-1} = \beta_2\beta_1^{-1} = \lambda \qquad N(\lambda) > 0$$

or $$\alpha_2\alpha_1^{1} = \beta_2\beta_1^{\prime -1} = \lambda \qquad N(\lambda) > 0$$

This is because by assumption

$$N(\alpha_2\alpha_1^{-1}) = N(\beta_2\beta_1^{-1}) > 0$$

$$tr(\alpha_2\beta_2') = N(\alpha_2\alpha_1^{-1})tr(\alpha_1\beta_1')$$

Note also that the form f is definite if $D < 0$, but indefinite if $D > 0$. In case $D < 0$ the behavior of f and of $-f$ can never be the same, whereas it is quite well possible that an indefinite binary form f is equivalent to $-f$; e.g., $f(x,y) = x^2 - 2y^2$ is equivalent to $-f(x,y)$.

Furthermore, let us define the *order of a Dedekind module* M to be the multiplier set

$$[M/M] = \{\lambda | \lambda \in K \text{ and } \lambda M \subseteq M\} \tag{39}$$

It is both a Dedekind module and a unital ring; in short, it is an *order* of K, following Dedekind's usage of the term.

It is known from the arithmetics of quadratic fields (see also Gauss [1], Sec. 306) that the orders of K are obtained in the form

$$\Omega_n = \mathbb{Z}1 \dotplus \mathbb{Z}n\omega \qquad n = 1, \ldots$$

where

$$\omega = \frac{D_0 + \sqrt{D_0}}{2}$$

and either

$$\begin{aligned} &D = k^2 D_0 \qquad 0 < k \in \mathbb{Z} \\ &D_0 \text{ square-free} \\ &D_0 \equiv 1 \pmod 4 \end{aligned} \tag{40c}$$

or

$$\begin{aligned} &4D = k^2 D_0 \qquad 0 < k \in \mathbb{Z} \\ &D_0 \equiv 0 \pmod 4 \\ &D_0/4 \text{ square-free} \end{aligned} \tag{40d}$$

Furthermore, it is known that the Dedekind modules with a fixed order Ω of K form an abelian group $I(\Omega)$ for the customary multiplication of submodules of a ring, the group of the invertible Ω ideals. The unit element of $I(\Omega)$ is Ω. The abelian group $I(\Omega)$ contains the subgroup $I_0(\Omega)$ of the proper principal ideals $\lambda\Omega$ with λ of K and $N(\lambda) > 0$.

The congruence relation of the ideals of $I(\Omega)$ modulo the normal subgroup $I_0(\Omega)$ also is called the *proper equivalence* of the invertible Ω-ideals. In other words, two Dedekind modules M_1, M_2 of K are properly equivalent precisely if there is an element λ of K of positive norm for which $M_2 = \lambda M_1$.

The proper equivalence classes contained in $I(\Omega)$ form the abelian group $I(\Omega)/I_0(\Omega)$ which is called the proper *Ω-ideal class group*.

THEOREM 3. The proper ideal class group of any Dedekind order Ω of the quadratic number field K is finite.

Proof. Let $\Omega = \Omega_n$. By construction $\Omega_n' = \Omega_n$, hence $\Omega' = \Omega$. It follows for any member M of $I(\Omega)$ also the algebraic conjugate module, M', belongs to Ω.

Because of the invertibility of M for any nonzero element λ of M there is a nonzero μ such that $M = \lambda\Omega + \mu\Omega$.

If $M \subseteq \Omega$ then $\Omega:M$ is a natural number, called the *norm of* M *in* Ω:

$$N(M) = \Omega:M \tag{41a}$$

For any principal ideal $\lambda\Omega$ we find by application of λ to a $\mathbb{Z}$-basis of Ω and evaluation of the determinant that

$$N(\lambda\dot{\Omega}) = \Omega:\ \lambda M = |N(\lambda)| \tag{41b}$$

If $\lambda \in M$, then $\lambda\Omega \subseteq M$, so that $\Omega:\ M|N(\lambda)$, $N(\lambda) = \lambda\lambda' \in \lambda M$. If $0 \neq \lambda$, then there is an element μ of M for which $N(\lambda)\Omega + \mu\Omega = M$. Hence $MM' = N(\lambda)^2\Omega + N(\lambda)\mu\Omega + N(\lambda)\mu'\Omega + N(\mu)\Omega \subseteq N(\lambda)\Omega + N(\mu)\Omega \subseteq MM'$,

$$MM' = N(\lambda)\Omega + N(\mu)\Omega$$

$$MM' = \gcd(N(\lambda),N(\mu))\Omega \tag{41c}$$

$$MM' = N(M) \tag{41d}$$

Hence, for M_1, M_2 of $I(\Omega)$, $M_1 \subseteq \Omega$, $M_2 \subseteq \Omega$ we have

$$N(M_1M_2) = M_1M_2(M_1M_2)' = M_1M_1'M_2M_2' = N(M_1)N(M_2)$$

$$N(M_1M_2) = N(M_1)N(M_2) \tag{41e}$$

The multiplicativity of the norm for the integral members of the ideal group $I(\Omega)$ remains in force for all members if we extend the norm definition uniquely to $I(\Omega)$ by setting

$$N(\lambda M) = |N(\lambda)|N(M) \qquad 0 \neq \lambda \in K,\ M \in I(\Omega),\ M \subseteq \Omega \tag{41f}$$

We apply these general remarks to the norm forms. By general principles

$$d(g_{\alpha s_{11}+\beta s_{12},\alpha s_{21}+\beta s_{22}}) = \det(s_{ik})^2 d(g_{\alpha,\beta})(\alpha,\beta \in K = 2\alpha + 2\beta) \tag{41g}$$

(where $d(g) = \det A_g$ for rational forms g) and for any 2×2 matrix (s_{ik}) with rational coefficients.

Hence the discriminant

$$d(\mathbb{Z}\alpha \dotplus \mathbb{Z}\beta/\mathbb{Z}) = |d(g_{\alpha,\beta})| \tag{41h}$$

depends only on the module $M = \mathbb{Z}\alpha \dotplus \mathbb{Z}\beta$, not on the choice of the **Z**-basis α, β of M. And if $M_2 \subseteq M_1$ then

$$d(M_2/\mathbb{Z}) = (M_1:M_2)^2 d(M_1/\mathbb{Z}) \tag{41i}$$

This relation remains in force for any two Dedekind submodules M_1, M_2 of K if we define

$$M_1:M_2 = (M_1: \ (M_1 \cap M_2))(M_2: \ (M_1 \cap M_2))^{-1} \tag{41j}$$

Note that the index multiplication relation

$$(M_1:M_2) = (M_2:M_3) = (M_1:M_3) \tag{41k}$$

also holds for any three Dedekind submodules M_1, M_2, M_3 of K.

Note that Ω_1 is the maximal order of K and that

$$d(\Omega_1/\mathbb{Z}) = |D_0| \tag{41ℓ}$$

The number D_0 also is called the *discriminant* of K over 2. Note that

$$\begin{aligned} \langle\Omega_n, \Omega_m\rangle &= \Omega_{\gcd(n,m)} \\ \Omega_n \cap \Omega_m &= \Omega_{\mathrm{lcm}(n,m)} \end{aligned} \qquad n,m \in \mathbb{Z} > 0 \tag{41m}$$

$$d(\Omega_n/\mathbb{Z}) = |D_0|n^2 \qquad 0 < n \in \mathbb{Z} \tag{41n}$$

For any member M of $I(\Omega_n)$ we have

$$d(M/\mathbb{Z}) = N(M)^2 d(\Omega_n/\mathbb{Z}) = N(M)^2 |D_0| n^2 \tag{41o}$$

We observe that the norm form of Ω_n has discriminant $D_0 n^2/4$. If $D_0 \equiv 0 \pmod 4$ then it is primitive, if $D_0 \equiv 1$ then 2 times the norm form is primitive. This remains true for the product of $N(M)^{-1}$ into the norm form defined on M, as we have shown above.

We know that there are only finitely many nonequivalent binary forms of any given nonzero discriminant according to Theorem 2. Hence there are only finitely many distinct equivalence classes represented by the products of $N(M)^{-1}$ into the norm form on M ($M \in I(\Omega_n)$). Because of the correspondence established above we infer the finiteness of the proper ideal class group of Ω_n.

THEOREM 4. The number of the proper equivalence classes of the primitive forms of discriminant $D = n^2 D_0$ which are derived from the norm forms of the members of $I(\Omega_n)$ is equal to the number of the proper ideal classes of Ω_n.

Proof. Note that the algebraic conjugate of a proper principal ideal $\lambda\Omega_n$ is the proper principal ideal.

$$(\lambda\Omega_n)' = \lambda'\Omega_n \tag{42a}$$

Hence the involutory automorphism $\alpha \to \alpha'$ of K induces an involutory automorphism on the proper ideal class group of Ω_n. A proper ideal class of Ω_n is said to be *ambiguous* if it coincides with its algebraic conjugate.

If M is a member of an ambiguous class then

$$M' = \lambda M \qquad \lambda \in K,\ N(\lambda) > 0$$

$$M = (M')'$$

$$M = N(\lambda)M$$

By a famous theorem of Hilbert there exists a nonzero element μ of K such that

$$\lambda = \frac{\mu}{\mu'} \tag{42b}$$

hence

$$(\mu M)' = \mu' M' = \mu'\lambda M = \mu M \tag{42c}$$

(A straightforward proof of (42b) is easy!) Thus it is shown that the ambiguous ideal class represented by M contains a member that is equal to its algebraic conjugate, i.e., an *ambiguous ideal*. The primitive forms associated with an ambiguous ideal are improperly equivalent to themselves. In this case there is only one class of forms and only one class of ideals.

If, however, the Ω-ideal M does not belong to an ambiguous ideal class, then the primitive forms associated with M, M' fall into two proper equivalence classes, as do the proper ideal classes associated with M, M'.

Thus Theorem 4 is demonstrated.

COMPOSITION OF FORMS

If $A = (a_{ik})$, $B = (b_{ik})$ are two symmetric integral matrices of degree m, n, respectively, then their *Kronecker product*

$$A \times B = (a_{i_1k_1} b_{i_2k_2}) \tag{43a}$$

is a symmetric integral matrix of degree mn when the row indices on the mn ordered couples (i_1, i_2) and the column indices are the mn ordered couples (k_1, k_2). In principle it does not matter which ordering of the row or column indices is adopted, all that matters is that it is the same for both rows and columns. Of the (mn)! possible orderings we usually adopt the lexicographic ordering

$$\begin{aligned}(i_1, i_2) < (j_1, j_2) &\Leftrightarrow i_1 < j_1 \quad \text{or} \quad i_1 = j_1, i_2 < j_2\\ &\Leftrightarrow i_1 m + i_2 < j_1 m + j_2\end{aligned}$$

One also adopts the Kronecker product definitions (43a) for rectangular matrices so that (i_1, i_2) ranges over the $m_1 n_1$ ordered couples (i_1, i_2) $(1 \le i_1 \le m_1,\ 1 \le i_2 \le n_1)$ and (k_1, k_2) ranges over the $m_2 n_2$ ordered couples (k_1, k_2) $(1 \le k_1 \le m_2,\ 1 \le k_2 \le n_2)$.

It is well known that the Kronecker product satisfies the distributivity rules

$$\begin{aligned}(A_1 + A_2) \times B &= A_1 \times B + A_2 \times B\\ A \times (B_1 + B_2) &= A \times B_1 + A \times B_2\end{aligned} \tag{43c}$$

the multiplication rule

$$(A_1 \times B_2)(C_1 \times C_2) = (A_1 C_1) \times (B_2 C_2) \tag{43d}$$

the identity law

$$I_m \times I_n = I_{mn} \tag{43e}$$

the inversion rule

$$(A \times B)^{-1} = A^{-1} \times B^{-1} \tag{43f}$$

the zero rule

$$0^{m_1 \times n_1} \times B_1 = A_1 \times 0^{n_2 \times n_2} = 0^{m_1 n_1 \times m_2 m_2} \tag{43g}$$

the negation rule

$$(-A_1) \times B_1 = A_1 \times (-B_1) = -(A_1 \times B_1) \tag{43h}$$

the scalar rule

$$(\lambda A_1) \times (\mu B_1) = \lambda\mu(A_1 \times B_1) \tag{43i}$$

and the transposition rule

$$(A_1 \times B_1)^T = A_1^T \times B_1^T \tag{43j}$$

($A_1, A_2 \in R^{m_1 \times n_1}$; $B_1, B_2 \in R^{m_2 \times n_2}$; $C_1 \in R^{n_1 \times p_1}$; $C_2 \in R^{n_2 \times p_2}$; $A \in GL(m,R)$; $B \in GL(n,R)$; $\lambda, \mu \in R$; and R a unital commutative ring.) Furthermore, there is a certain permutation matrix of degree mn mapping the lexicographically ordered index couples for $A_1 \times B_1$ into the ones for $B_1 \times A_1$. Finally

$$(A_1 \times B_1) \times C_1 = A_1 \times (B_1 \times C_1) \tag{43k}$$

($A_1 \in R^{m \times m}$, $B_1 \in R^{n \times n}$, $C_1 \in R^{p \times p}$, R any ring). Correspondingly, we associate with any two forms, say with the m-ary quadratic form f and the n-ary quadratic form g, the *Kronecker product form* $f \times g$ defined by

$$A_{f \times g} = A_f \times A_g \tag{43ℓ}$$

We observe that the matrix equations

$$S_1 A_f S_1^T = A_{f_1} \qquad S_2 A_g S_2^T = A_{g_1} \qquad S_1 \in \mathbb{Z}^{p \times m},\ S_2 \in \mathbb{Z}^{q \times n} \tag{43m}$$

imply the matrix equation

$$(S_1 \times S_2) A_f \times A_g (S_1 \times S_2)^T = A_{f_1} \times A_{g_1} \tag{43n}$$

hence

$$f \geq f_1,\ g \geq g_1 \Rightarrow f \times g \geq f \times g_1 \tag{43o}$$

in particular

$$f \underset{\mathbb{Z}}{\sim} f_1,\ g \underset{\mathbb{Z}}{\sim} g_1 \Rightarrow f \times g \underset{\mathbb{Z}}{\sim} f_1 \times g_1$$

so that the Kronecker multiplication of the arithmetic equivalence class is uniquely defined by

$$(f) \times (g) = (f \times g) \tag{43q}$$

According to the observations made above regarding the commutativity and associativity of Kronecker multiplications of matrices, it follows that the Kronecker multiplication of the arithmetic equivalence classes of quadratic forms is commutative and associative. Note that the form

$$\mathbb{1}(x) = x^2 \qquad A_{\mathbb{1}} = (1) \tag{43r}$$

of degree 1 represents the neutral element inasmuch as

$$(1) \times B_1 = B_1 \times (1) = B_1 \qquad (B_1 \in R^{n\times n},\ R \text{ a unital ring}) \tag{43s}$$

$$(\mathbb{1}) \times (f) = (f) \times (\mathbb{1}) = (f) \qquad (f \text{ any n-ary form}) \tag{43t}$$

The Kronecker multiplication of matrices and forms is reflected by the Kronecker multiplication of modules and lattices.

Suppose L_j is an r-dimensional vector lattice contained in $\mathbb{R}^{1\times n_j}$ $(j = 1, 2)$, then the Kronecker product of L_1, L_2 is defined as the r_1r_2-dimensional vector lattice:

$$L_1 \times L_2 = \{x_1 \times x_2 | x_j \in L_j,\ j = 1, 2\} \tag{44a}$$

contained in $\mathbb{R}^{1\times(n_1n_2)}$.

Suppose there is the Z-basis $b_{h_r}, \ldots, b_{r_j}$ of L_j and the symmetric matrix X_j of $\mathbb{R}^{n_j\times n_j}$ such that

$$(b_{ji}X_jb_{jk}^T) = (a_{ik}) = A_{f_j} \in \mathbb{Z}^{r_j\times r_j} \qquad j = 1, 2 \tag{44b}$$

then the r_1r_2 elements $b_{1i} \times b_{2k}$ form a $\mathbb{Z}$-basis of $L_1 \times L_2$ such that

$$((b_{1i_1} \times b_{2i_2})X_1 \times X_2(b_{1k_1} \times b_{2k_2})^T) = A_{f_1\times f_2} \tag{44c}$$

More generally, let R be a unital commutative ring and let M_1, M_2 be two unital R-modules, then their *R-tensor product* is defined as the unital R-module $M_1 \underset{R}{\times} M_2$ with the generators $u_1 \times u_2$ $(u_j \in M_j,\ j = 1, 2)$, the defining relations

$$\begin{aligned}(u_1 + v_1) \times u_2 &= u_1 \times u_2 + v_1 \times u_2 \quad (u_j \in M_j,\ v_j \in M_j,\ j = 1, 2)\\ u_1 \times (u_2 + v_2) &= u_1 \times u_2 + u_1 \times v_2\end{aligned} \tag{44d}$$

$$(\lambda u) \times v = u \times (\lambda v) \qquad \lambda \in R,\ u \in M_1,\ v \in M_2 \tag{44e}$$

and the scalar multiplication rule

$$\lambda\left(\sum_{i=1}^{s} z_i \times z_i\right) = \sum_{i=1}^{s} (\lambda z_i) \times z_i \tag{44f}$$

$$(z_i \in M_1,\ z_i \in M_2,\ 1 \le i \le s,\ \lambda \in R)$$

For example, the Kronecker product of two lattices is isomorphic to their $\mathbb{Z}$-tensor product

$$L_1 \underset{\mathbb{Z}}{\times} L_2 \sim L_1 \underset{\mathbb{Z}}{\otimes} L_2 \tag{44g}$$

For any unital R-module M_1 the R-bilinear mappings of M_1 in R are the R-linear mappings of $M_1 \underset{R}{\otimes} M_1$ in R.

If M_1, M_2 are any two unital R-modules and if σ_j is an R-linear mapping of M_j into the unital R-module N_j, then there is the R-linear mapping

$$M_1 \underset{R}{\otimes} M_2 \to N_1 \underset{R}{\otimes} N_2 :\ \sigma_1 \underset{R}{\otimes} \sigma_2 \tag{44h}$$

for which

$$(u_1 \otimes u_2)\sigma_1 \underset{R}{\otimes} \sigma_2 = u_1\sigma_1 \otimes u_2\sigma_2 \tag{44i}$$

Suppose τ_j is a bilinear mapping on M_j into R ($j = 1, 2$), then $\tau_1 \underset{R}{\otimes} \tau_2$ is a bilinear mapping on $M_1 \underset{R}{\otimes} M_2$ into R. If both τ_1, τ_2 are symmetric, i.e., if

$$(u_1 \otimes u_2)\tau_j = (u_2 \otimes u_1)\tau_j \qquad u_1, u_2 \in M_j \qquad j = 1, 2 \tag{44j}$$

then also $\tau_1 \underset{R}{\otimes} \tau_2$ is symmetric. In this way the Kronecker product of forms is defined even if there is no basis of the underlying module over the coefficient ring.

Gauss (in [1], article 235 and subsequently) treats the problem of the composition of binary forms in very explicit terms. Let us take up the task in a more general way and let us make full use of the conventions of matrix calculus.

A *quadratic space* over an integral domain R is defined as a finitely generated torsion-free R-module V together with a symmetric bilinear form

$$V \underset{F}{\otimes} V \to R : \beta \tag{45a}$$

If there is an R-basis $b_1, \ldots, b_r$ of V, then β is associated with respect to the R-basis $b_1, \ldots, b_r$ of V with the symmetric matrix

$$A_{\beta;b_1,\ldots,b_r} = ((b_i \otimes b_k)\beta) \tag{45b}$$

and conversely.

Note that each R-quadratic space V is uniquely embedded into the quadratic space $\bar{V} = Q(R) \underset{R}{\otimes} V$ over the quotient field Q(R) of R with symmetric bilinear form

$$\bar{V} \underset{Q(R)}{\otimes} \bar{V} \to Q(R) : \bar{\beta} \tag{45c}$$

$$((\lambda \underset{R}{\otimes} u) \underset{Q(R)}{\otimes} (\mu \underset{R}{\otimes} v))\beta = (\lambda\mu) \underset{R}{\otimes} ((u \underset{R}{\otimes} v)\beta) \tag{45d}$$

$$(\lambda, \mu \in Q(R);\ u, v \in V)$$

The R-rank of V is equal to the Q(R)-dimension of $\bar{V}$.

For any subset S of the quadratic R-space V its *orthogonal complement* is defined as the subset

$$S^{\perp} = \{x | x \in V \text{ and } \forall\, s(s \in S \Rightarrow (s \underset{R}{\otimes} x) = 0)\} \tag{46a}$$

The subset is a purely embedded R-invariant submodule of V, in other words,

$$(Q(R) \underset{R}{\otimes} S^{\perp}) \cap V = S^{\perp} \tag{46c}$$

Furthermore:

$$S^{\perp} = (RS)^{\perp} = (S + S)^{\perp} \tag{46c}$$

$$(S_1 \cup S_2)^{\perp} = ((S_1 \cup 0) + S_2)^{\perp} = S_1^{\perp} \cap S_2^{\perp} \tag{46d}$$

$$(S_1 \cap S_2)^{\perp} = (Q(R) \underset{R}{\otimes} (S_1^{\perp} + S_2^{\perp})) \cap V \tag{46e}$$

$$((S^{\perp})^{\perp})^{\perp} = S^{\perp} \tag{46f}$$

$(S^{\perp})^{\perp} = S$ precisely if S is an R invariant submodule purely embedded into V and containing $V^{\perp}$ (46g)

$$\text{rank } S^{\perp} + \text{rank } (S^{\perp})^{\perp} = \text{rank } V + \text{rank } V^{\perp} \tag{46h}$$

The quadratic space V is said to be *nondegenerate* if $V^{\perp} = 0$.

For any extension F of Q(R) we embed the quadratic R-space V into the quadratic F-space $F \underset{R}{\otimes} V$, subject to

$$((\lambda \underset{R}{\otimes} u) \underset{F}{\otimes} (\mu \otimes v))\bar{\beta} = (\lambda\mu) \underset{R}{\otimes} ((u \underset{R}{\otimes} v)\beta) \tag{46i}$$

$$(\lambda,\ \mu \in F;\ u,\ v \in V)$$

such that for any subset S of V the S perp in $F \underset{R}{\otimes} V$ is equal to $F \underset{R}{\otimes} S^{\perp}$. The quadratic F-space $F \underset{R}{\otimes} V$ is nondegenerate precisely if the quadratic R-space V is nondegenerate.

The quadratic R-space V of rank r is nondegenerate precisely if there are 2r elements $b_1, \ldots, b_r, c_1, \ldots, c_r$ of V for which $\det((b_i \times c_k)\beta) \neq 0$. If this nonequation holds, then both the r-tuple $b_1, \ldots, b_r$ as well as the r-tuple $c_1, \ldots, c_r$ are linearly independent over R.

The R-ideal generated by all determinants

$$\det((b_i \times c_k)\beta) \text{ with } b_1, \ldots, b_r, c_1, \ldots, c_r$$

of V is said to be the *discriminant ideal* of the quadratic space over R. It is nonzero precisely if V is nondegenerate. If there is an R-basis $a_1, \ldots, a_r$ of V, then the discriminant ideal of V over R is the principal ideal generated by $\det((a_i \times a_k)\beta)$.

The latter expression is said to be a *discriminant* of V over R. It is uniquely determined up to the square of a unit factor.

Given two quadratic R-spaces V_j such that

$$V_j \underset{R}{\otimes} V_j \to R : \beta_j \qquad j = 1,\ 2 \tag{47}$$

and that V_1, V_2 both have the same R-rank r then we speak of the quadratic R-space V of R-rank r, and with symmetric bilinear R-mapping (45a) as a *composite* of V_1, V_2 if there is an R-epimorphism

$$V_1 \underset{R}{\otimes} V_2 \to V : \varepsilon \tag{48a}$$

such that

$$((\mu_1 \otimes u_2)\varepsilon \otimes (u_1 \otimes u_2)\varepsilon)\beta = ((u_1 \otimes u_1)\beta_1)((u_2 \otimes u_2)\beta_2) \tag{48b}$$

$$(u_j \in V_j;\ j = 1,\ 2)$$

For which values of r does there exist a composition? Theorem 3 (Hurwitz [1]): If the characteristic of R is not 2, then r = 1, 2, 4, or 8.

Both assumptions and assertions are proved upon replacement of R by an algebraically closed extension of Q(R). Without loss of generality R may be assumed to be an algebraically closed field F of characteristic $\neq 2$.

Under that assumption it follows that there are bases

$$b_{ji} \text{ of } V_j \quad (j = 1, 2) \qquad b_i \text{ of } V \quad (i = 1, 2, \ldots, r)$$

such that

$$(b_{ji} \otimes b_{jk})\beta_j = (b_i \otimes b_k)\beta = \delta_{ik} \quad (i, k = 1, 2, \ldots, r) \tag{48c}$$

The question is for which values of r an identity of the form

$$\sum_{k=1}^{r} z_k^2 = \sum_{i=1}^{r} x_i^2 \sum_{j=1}^{r} y_j^2 \tag{48d}$$

for 2r independent variables $x_1, \ldots, x_r, y_1, y_2, \ldots, y_r$ and with bilinear forms

$$z_k = \sum_{i=1}^{r} \sum_{j=1}^{r} \alpha_{ijk} \times x_i y_j \tag{48e}$$

over F can hold.

The identity is tantamount to the relations

$$\sum_{k=1}^{r} (\alpha_{ijk}\alpha_{i'j'k} + \alpha_{ij'k}\alpha_{i'jk}) = 2\delta_{ii'}\delta_{jj'} \tag{48f}$$

$$(i,i', j,j' = 1, 2, \ldots, r)$$

which turns into the matrix equations

$$A_i A_{i'}^T + A_{i'} A_i^T = 2\delta_{ii'} I_r \tag{48g}$$

or, equivalently,

$$A_r \in GL(r,F) \qquad A_r^T = A_r^{-1} \tag{48h}$$

$$B_i = A_i A_r^{-1} \in GL(r,F) \qquad A_i = B_i A_r \qquad 1 \le i < r \tag{48i}$$

$$B_i^T = -B_i = B_i^{-1} \qquad 1 \le i < r \tag{48j}$$

$$(B_j B_k)^2 = -I_r \qquad 1 \le j < k < r \tag{48k}$$

A look at the relations (48j), (48k) suggests forming the abstract group G with r - 1 generators $g_1, \ldots, g_{r-1}$ and defining relations

$$g_i^2 = \varepsilon \qquad 1 \le i < r \tag{48ℓ}$$

$$(g_j g_k)^2 = \varepsilon \qquad 1 \le j < k < r \tag{48m}$$

with the representation ψ of degree r over F mapping g_i on B_i $(1 \le i < r)$, ε on $-I_r$.

$r = 1$: $A_1 = \pm 1$ satisfies (48h).
$r = 2$: $A_1 = \pm I_2$, $B_i = \pm\begin{pmatrix} 0 & 1 \\ -1 & 0 \end{pmatrix}$ satisfies (48i)-(48k).
$r > 2$: Note that (48m) is tantamount to

$$\varepsilon^2 = 1 \qquad [g_j, g_k] = g_j^{-1} g_k^{-1} g_j g_k = \varepsilon \qquad 1 \le j < k < r \tag{48n}$$

The order of G is 2^r with $\langle\varepsilon\rangle$ as commutator group of order 2.

If r is odd then $\langle\varepsilon\rangle$ is the center, the classes of conjugate elements of G are the two center elements together with the $2^{r-1} - 1$ noncentral cosets modulo $\langle\varepsilon\rangle$. Hence there are precisely $2^{r-1} + 1$ nonequivalent proper irreducible representations of which 2^{r-1} are of degree 1, mapping ε on 1. The remaining irreducible representation is of degree $2^{(r-1)/2}$ in harmony with Frobenius' degree equation

$$2^r = 2^{r-1} \times 1^2 + 1 \times (2^{(r-1)/2})^2$$

It maps ε on $-I_{2^{(r-1)/2}}$, and ψ is a multiple, hence

$$2^{(r-1)/2} | r$$

which is impossible because r is odd. If r is even and ≥ 4 then the center of G is $\langle\varepsilon \times g_1 \cdots g_{r-1}\rangle$, the classes of conjugate elements of G are are the four center elements together with the $2^{r-1} + 2$ noncentral cosets modulo $\langle\varepsilon\rangle$. Hence there are precisely $2^{r-1} = 2$ nonequivalent proper irreducible representations of which 2^{r-1} are of degree 1, mapping ε on 1. The remaining two irreducible representations each are of degree $2^{(r-2)/2}$ in harmony with Frobenius' degree equation

$$2^r = 2^{r-1} \times 1^2 + 2 \times (2^{(r-2)/2})^2$$

and each of them map ε on $-I_{2^{(r-2)/2}}$.

The representation ψ has only irreducible constituents of degree $2^{(r-2)/2}$ whence

$$2^{(r-2)/2} | r$$

hence $r = 4, 8$.

If r = 4 then there are two irreducible representations of degree 2, say ψ_1, ψ_2, hence $\psi \sim 2\psi_j$ (j = 1, 2) or $\psi \sim \psi_1 \dotplus \psi_2$, but the latter alternative is impossible because all matrices X solving the condition

$$x(\psi_1 \dotplus \psi_2)X = Xx(\psi_1 \dotplus \psi_2)^{-T} \qquad x \in g$$

are skew-symmetric so that $g(\psi_1 \dotplus \psi_2)$ cannot be transformed into a subgroup of O(4,F).

Hence, $\psi = \dot{2}\psi_1$, say, and $g\psi$ is a quaternion group

$$A_4 = I_4, \quad B_1\psi = \begin{vmatrix} 0 & -1 & & \\ 1 & 0 & & \\ & & 0 & -1 \\ & & 1 & 0 \end{vmatrix}, \quad B_2\psi = \begin{vmatrix} & & -1 & 0 \\ & & 0 & 1 \\ 1 & 0 & & \\ 0 & -2 & & \end{vmatrix},$$

$$B_3\psi = (B_1\psi)(B_2\psi) \tag{48p}$$

If r = 8 then ψ is irreducible of degree 8.

COMPUTATIONS OF THE ZERO PRESENTATION FOR TERNARY FORMS

Every homogeneous polynomial f in n variables $x_1, \ldots, x_n$ over an integral domain I presents zero trivially according to

$$f(0,\ldots,0) = 0 \tag{1}$$

Any zero presentation

$$f(\xi_1,\ldots,\xi_n) = 0 \qquad \xi_1, \ldots, \xi_n \in I \tag{2}$$

implies the zero presentation

$$f(\lambda\xi_1,\ldots,\lambda\xi_n) = 0 \tag{3}$$

for arbitrary elements λ of I, and vice versa.

The zero presentation (2) is said to be *primitive* if any common divisor of $\xi_1, \ldots, \xi_n$ is a unit of I. Primitive zero presentations are nontrivial. If I satisfies the *ascending divisor chain condition* (I is an ADC-domain): $\lambda_i \in I$, $\lambda_{i+1}|\lambda_i$ ($i \in \mathbf{Z} > 0$) implies that $\lambda_j|\lambda_{j+1}$ for some index j' then any nontrivial zero presentation is obtained from a primitive zero presentation in the same way (3) is derived from (2).

Similarly, if f is not identically zero then f = μ with nonzero element μ of the ACD-domain I such that the coefficients of the homogeneous polynomial g in $x_1, \ldots, x_n$ have only units of I as their common divisor. We say f *is a scalar multiple of the primitive polynomial* g and that g is uniquely determined by f up to a unit factor.

Equation (2) is tantamount to

$$g(\xi_1,\ldots,\xi_n) = 0 \tag{4}$$

The nontrivial zero presentations of f are derived from the primitive zero presentations (4) by g in the form (3). More generally, any nonzero homogeneous polynomial f with coefficients in the quotient field Q(I) of I is of the form $f = \mu g$, where g is a primitive homogeneous polynomial in $x_1, \ldots, x_n$ over I, and μ is a nonzero element of Q(I). Again, the nontrivial zero presentations of f over Q(I) are derived from the primitive zero presentations (4) by g over I in the form (3) with λ a nonzero element of Q(I).

We note that for any nonsingular linear substitution $x_i = \Sigma_{i=1}^{n}\alpha_{ik}y_k$, $(\alpha_{ik}) \in GL/n,Q(I)$, the polynomial $\hat{f}$ of $y_1, \ldots, y_n$ defined by setting

$$\hat{f}(y_1,\ldots,y_n) = f(x_1,\ldots,x_n) \tag{5}$$

is homogeneous of the same degree in $y_1, \ldots, y_n$ as f is in $x_1, \ldots, x_n$. Moreover, any zero presentation (2) by f leads to the zero presentation

$$\hat{f}(\eta_1,\ldots,\eta_n) = 0 \tag{5a}$$

with

$$\eta_i = \sum_{k=1}^{n} \alpha_{ik}\xi_k \qquad i = 1, 2, \ldots, n \tag{5b}$$

and vice versa.

For the ADC-domain $\mathbb{Z}$ any nonzero quadratic form f is equivalent to a nondegenerate quadratic form g such that

$$A_g = XA_fX^T \tag{6a}$$

with a certain rectangular matrix X of $\mathbb{Z}^{m\times n}$. Any zero permutation

$$f(x) = xA_fx^T = 0 \qquad x \in \mathbb{Z}^{1\times n} \tag{6b}$$

by f leads to a zero presentation

$$g(y) = yA_gy^T = 0 \qquad y \in \mathbb{Z}^{1\times m}$$

by g such that

$$yX = x + u \tag{6d}$$

where the n-row u over $\mathbb{Z}$ satisfies the condition

$$uA_f = 0 \tag{6e}$$

and vice versa.

Thus the problem of studying zero presentations by ternary forms is reduced to the study of the primitive zero presentations by primitive nondegenerate ternary diagonal forms

$$f = \alpha_1 x_1^2 + \alpha_2 x_2^2 + \alpha_3 x_3^2 \tag{7a}$$

where

$$\alpha_i \in \mathbb{Z}\ (1 \le i \le 3),\ d(f) = \alpha_1\alpha_2\alpha_3 > 0,\ \gcd(\alpha_1,\alpha_2,\alpha_3) = 1 \tag{7b}$$

It is clear that a definite form only has the trivial zero presentation.

There remains the case that the three nonzero rational integers are not all of the same sign.

I. $d(f) = 1$

There is just one equivalence class represented by

$$f = -x_1^2 + x_2^2 - x_3^2 \tag{7c}$$

with the primitive zero presentation $f(1,1,0) = 0$. The Gauss reduction method shows how to construct a unimodular matrix transforming f into normal form (7c).

What are the other primitive zero presentations?

Let us show that they are all equivalent to the standard zero presentation. Indeed, let $\xi = (\xi_1,\xi_2,\xi_3)$ be a primitive zero presentation.

All elements η of $\mathbb{Z}^{1\times 3}$ satisfying the condition $\xi A_f \eta^T = 0$ form a submodule ξ of $\mathbb{Z}^{1\times 3}$ containing ξ, of rank 2 and purely embedded into $\mathbb{Z}^{1\times 3}$. Since the zero presentation $\xi A_f \xi^T = 0$ is primitive, it follows that there is a basis $\xi = \xi^{(1)}, \xi^{(2)}, \xi^{(3)}$ of $\mathbb{Z}^{1\times 3}$ such that $\xi^1 = \mathbb{Z}\xi^{(1)} + \mathbb{Z}\xi^{(2)}$. Hence

$$1 = \det(f) = \det(\xi^{(i)} A_f \xi^{(k)T})$$

$$= \det \begin{pmatrix} 0 & 0 & \xi^{(1)} A_f \xi^{(3)T} \\ 0 & \xi^{(2)} A_f \xi^{(2)T} & * \\ \xi^{(3)} A_f \xi^{(1)T} & * & * \end{pmatrix}$$

$$= -\xi^{(2)} A_f \xi^{(2)T} (\xi^{(1)} A_f \xi^{(3)T})$$

$$\xi^{(2)} A_f \xi^{(2)T} = -1 = \pm \xi^{(1)} A_f \xi^{(3)T}$$

upon replacement of $\xi^{(3)}$ by $\xi^{(3)} + \lambda \xi^{(1)}$ for suitable integral λ $\xi^{(3)} A_f \xi^{(3)T} = 0$; upon replacement of $\xi^{(2)}$ by $\xi^{(2)} + \mu \xi^{(1)}$ for suitable integral μ we also have $\xi^{(2)} A_f \xi^{(3)T} = 0$. Replacing $\xi^{(3)}$ by $-\xi^{(3)}$, $\xi^{(2)}$ by $-\xi^{(2)}$ if necessary, we find that

$$(\xi^{(i)} A_f \xi^{(k)T}) = \begin{pmatrix} 0 & 0 & -1 \\ 0 & -1 & 0 \\ -1 & 0 & 0 \end{pmatrix}$$

and since the matrix on the right is independent of the choice of $\xi = \xi^{(1)}$, it follows indeed that all primitive solutions of (1) are equivalent.

II. $d(f) > 1$. We assume that the primitive zero presentations can be determined already up to equivalence in case the determinant is positive and less than $d(f)$.

IIa. Two of the coefficients $\alpha_1, \alpha_2, \alpha_3$, say α_1, α_3, are divisible by the same prime number p. It follows that $p \nmid \alpha_2$ and that for every primitive zero presentation

$$0 = f(\xi_1, \xi_2, \xi_3) = \alpha_1 \xi_1^2 + \alpha_2 \xi_2^2 + \alpha_3 \xi_3^2 \equiv \alpha_2 \xi_2^2 \pmod p)$$

$$\xi_2 \equiv 0 \pmod p), \quad \xi_2 = p\eta_2 \text{ with } \eta_2 \in \mathbb{Z}$$

Considering that $\alpha_1 = p\beta_1$, $\alpha_3 = p\beta_3$ with β_1, β_3 of $\mathbb{Z}$, it follows that the nontrivial zero presentation

$$0 = \alpha_1 \xi_1^2 + \alpha_2 \xi_2^2 + \alpha_3 \xi_3^2$$

by f is in one-to-one correspondence with the nontrivial zero presentation

$$0 = \beta_1\eta_1^2 + \beta_2\eta_2^2 + \beta_3\eta_3^2 \qquad (\beta_2 = \alpha_2,\ \eta_1 = \xi_1,\ \eta_3 = \xi_3)$$

by

$$g(x_1,x_2,x_3) = \beta_1 x^2 + \beta_2 x_2^2 + \beta_3 x_3^2$$

Since $0 < d(g) = \beta_1\beta_2\beta_3 = (1/p^2)d(f) < d(f)$, we have succeeded in reducing the problem to a case that could be settled already.

In this case the existence of a primitive zero presentation by f leads to congruence conditions modulo d(f) which enabled Gauss to explicitly construct an integral 3 × 3 matrix X for which

$$d(f)B = XA_f X^T \tag{8}$$

such that B is a symmetric integral matrix of determinant 1. According to I, we find the primitive zero presentations by the ternary form g corresponding to B which lead back via (8) to the zero presentation by f.

Gauss' method is based on the (nowadays) well-known observation of linear diophantine analysis that the natural homomorphism

$$\begin{aligned} &GL(n,\mathbb{Z}) \to GL(n,\mathbb{Z}/m):\ \varepsilon_m \\ &(\alpha_{ik})\varepsilon_m = (\alpha_{ik}/m\mathbb{Z}) \\ &(\alpha_{ik}) \in GL(n,\mathbb{Z}) \end{aligned} \tag{9a}$$

is an epimorphism such that for each image explicitly an original can be constructed. Furthermore, it uses the Chinese remainder theorem with its consequence

$$GL(n,\mathbb{Z}/m_1m_2) \simeq GL(n,\mathbb{Z}/m_1) \times GL/n,\mathbb{Z}/m_2) \tag{9b}$$

$$(m_1,\ m_2 \in \mathbb{Z},\ \gcd(m_1,m_2) = 1)$$

where the m-components of $(\alpha_{ik}/m_1m_2\mathbb{Z})$ are $(\alpha_{ik}/m_j\mathbb{Z})$, $(\alpha_{ik}) \in GL(n,\mathbb{Z})$; $j = 1,\ 2$.

If p is a prime number dividing α_j, say

$$p^\nu | \alpha_j \qquad p^{\nu+1} \nmid \alpha_j \qquad \nu > 0$$

and if there is a primitive solution of (1), then we have

$$\xi_h = p^{\nu_1}\eta_h \qquad \eta_h \in \mathbb{Z}\ (1 \le h \le 3,\ h \ne j) \text{ with } \nu_1 \in \mathbb{Z} \ge 0 \tag{10a}$$

$$p \nmid \eta_{h_1} \tag{10b}$$

where h_1 is one of the three numbers 1, 2, 3 distinct from j.

If $\nu_1 > 0$, then $p \nmid \xi$ because of the primitivity of the zero presentation (1); on the other hand

$$0 = \alpha_1\xi_1^2 + \alpha_2\xi_2^2 + \alpha_3\xi_3^2 \equiv \alpha_j\xi_j^2 \pmod{p^{2\nu_1}}$$

hence

$$2\nu_1 \le \nu,\ \alpha_j = p^{\nu_1}\beta_j,\ \beta_j \in \mathbb{Z}$$

so that for $\beta_h = \alpha_h$ $(1 \le h \le 3,\ h \ne j)$ we have the primitive zero presentations

$$0 = \beta_1\eta_1^2 + \beta_2\eta_2^2 + \beta_3\eta_3^2$$

by the ternary form

$$g(x_1,x_2,x_3) = B_1x_1^2 + B_2x_2^2 + B_3x_3^2$$

of positive determinant $d(f)/p^{\nu_1}$ less than d(f), and conversely. We can survey those primitive zero presentations by f already.

If $\nu_1 = 0$ then (1) implies that

$$p \nmid \xi_h \quad (1 \le h \le 3,\ h \ne j) \qquad \alpha \sum_{h \ne j} \alpha_h\xi_h^2 \equiv 0 \pmod{p^\nu}$$

Hence there is a matrix (α_{ik}) of $GL(3,\mathbb{Z})$ such that $f(x_1,x_2,x_3) = \hat{f}(y_1,y_2,y_3)$, where

$$x_i = \sum_{h=1}^{3} \sigma_{ik}y_k \qquad (\sigma_{ik}) \in GL(3,\mathbb{Z})$$

$$A_{\hat{f}} = (\beta_{ik}) \in \mathbb{Z}^{3\times 3}$$

$$\beta_{ik} \equiv 0 \pmod{p^\nu} \quad \text{if } i, k = 1, 2$$

Suppose we find $\nu_1 = 0$ for all prime numbers p dividing d(f), then it follows from the previous remarks that the unimodular matrix (σ_{ik}) can be determined in such a way that

$$\beta_{ik} \equiv 0 \pmod{d(f)} \quad \text{if } i, k = 1, 2, 3;\ i, k \ne 3, 3$$

It follows that

$$\hat{f}(y_1,y_2,d(f)y_3) = d(f)g(y_1,y_2,y_3)$$

where the ternary form g has determinant 1. Its coefficients are $\beta_{ik}/d(f)$ if i, k = 1, 2 or if i = k = 3 and β_{ik} otherwise. Thus the question is reduced to the discussion of the primitive zero presentations by a ternary form of determinant 1.

The following modification of Gauss' method permits us to develop the algorithm in a straightforward manner for any nondegenerate ternary form f.

If $d(f) < 0$, replace f by $-f$.

If $d(f) = 1$, we know already what to do.

Let $d(f) > 1$ and let p be a prime number dividing d(f). The congruence $d(f) \equiv 0 \pmod p$ indicates that the ternary form f over $\mathbb{Z}/p$, derived from f by replacing each coefficient of A_f by its residue class modulo p, is degenerate. We construct explicitly a unimodular matrix transforming f into a ternary form $\hat{f}$ of determinant $d(\hat{f}) = d(f)$ such that

$$A_{\hat{f}} = (\alpha_{ik}) \qquad \alpha_{ik} \in \mathbb{Z} \qquad i, k = 1, 2, 3$$

$$\alpha_{ik} \equiv \alpha_{k1} \equiv 0 \pmod p \qquad k = 1, 2, 3 \tag{11a}$$

The primitive zero presentations of f and g are in one-to-one correspondence. Without loss of generality we may assume that

$$A_f = (\alpha_{ik}) \tag{11b}$$

and that (11a) holds.

Among the primitive zero presentations (1) by f we distinguish two kinds.

a. $\xi_2 \equiv \xi_3 \equiv 0 \pmod p$

This implies that $\xi_1 \not\equiv 0 \pmod p$, $\alpha_{11} \equiv 0 \pmod{p^2}$. Setting

$$\alpha_{1k} = p\beta_{1k} = p\beta_{k1} \qquad k = 2, 3$$

$$\alpha_{11} = p^2\beta_{11}$$

$$\alpha_{ik} = \beta_{ik} \qquad i, k = 2, 3$$

$$\xi_k = p\eta_k \qquad k = 2, 3$$

$$\xi_1 = \eta$$

$$A_g = (\beta_{ik})$$

the primitive zero presentation (2) by f is associated with the primitive zero presentation

$$\Sigma \beta_{ik}\eta_i\eta_k = 0 \tag{11c}$$

by the ternary form g of determinant $d(f)/p^2$.

b. ξ_2, ξ_3 not both divisible by p.

In this case we can construct explicitly a unimodular matrix transforming f into a ternary form $\hat{f}$ of determinant $d(\hat{f}) = d(f)$ such that

$$A_{\hat{f}} = (\hat{\alpha}_{ik})$$

$$\hat{\alpha}_{ik} \quad \mathbb{Z} \quad (i, k = 1, 2, 3)$$

$$\hat{\alpha}_{ik} \equiv \hat{\alpha}_{k1} \equiv \hat{\alpha}_{22} \equiv 0 \pmod{p} \qquad (k = 1, 2, 3) \tag{11d}$$

and either

$$\hat{\alpha}_{3k} \equiv \hat{\alpha}_{k3} \equiv 0 \pmod{p} \qquad (k = 1, 2) \tag{11e}$$

$$\hat{\alpha}_{33} \not\equiv 0 \pmod{p}$$

or

$$p = 2 \qquad \hat{\alpha}_{33} \equiv \hat{\alpha}_{32} = \hat{\alpha}_{23} \equiv 0 \pmod{2} \tag{11f}$$

$$\hat{\alpha}_{13} = \hat{\alpha}_{31} \equiv 1 \pmod{2}$$

Without loss of generality we may assume that $f = \hat{f}$.

In the case of (11e) it follows that $\xi_3 \equiv 0 \pmod{p}$. Setting

$$\alpha_{ik} = p\beta_{ik} \qquad i, k = 1, 2$$

$$\alpha_{k3} = \alpha_{3k} \qquad k = 1, 2$$

$$p^2\alpha_{33} = \beta_{33}$$

$$\xi_k = \eta_k \qquad k = 1, 2$$

$$\xi_3 = p\eta_3$$

$$A_g = (\beta_{ik})$$

it follows that the primitive zero presentation (2) by f is associated with the primitive zero presentation (11c) by g, a ternary form of determinant $d(f)/p^2$.

In the case of (11f), $\alpha_{22} \equiv 0 \pmod 4$, set

$$\alpha_{ik} = \beta_{ik} \qquad i, k = 1, 2, 3;\ i, k \neq 2$$

$$\alpha_{2k} = 2\beta_{2k} = 2\beta_{k2} \qquad k = 1, 3$$

$$\alpha_{22} = 4\beta_{22}$$

$$\xi_k = \eta_k \qquad k = 1, 2$$

$$2\xi_k = \eta_2$$

$$A_g = (\beta_{ik})$$

It follows that the primitive zero presentation (1) by f is associated with the primitive zero presentation (11c) by g, a ternary form of determinant d(f)/4.

There remains the case (11f), $\alpha_{22} \equiv 2 \pmod 4$ in which d(f) is a power of 2. It follows that d(f) = 2. By Gauss' reduction method, we construct explicitly a unimodular matrix transforming f into its canonical form $f(x_1,x_2,x_3) = 2x_1x_3 - 2x_2^2$. It is clear that there are primitive zero presentations and how they are obtained: Set either

$$\xi_2 = \xi_3 = 0 \qquad \xi_1 = \pm 1$$

or $$\xi_2 = \xi_1 = 0 \qquad \xi_3 = \pm 1$$

or $$\xi_2 \neq 0 \qquad \xi_2 = -\xi_1'\xi_3' \qquad \gcd(\xi_1', \xi_3') = 1$$

$$\xi_1 = \xi_1'^2 \qquad \xi_3 = \xi_3'^2$$

Gauss' analysis as well as its modification reveals the local-global statement according to which a nontrivial zero presentation by the ternary form f exists if and only if it exists for its residuation modulo an arbitrary prime power or modulo infinity (i.e., f is not definite). It suffices to study the behavior of f modulo the prime powers occurring in the factorization of the discriminant.

THE AUTOMORPH OF A DEGENERATE FORM

THEOREM. The $\mathbb{Z}$-automorph of a nondegenerate form f is a finitely presentable discrete subgroup of its $\mathbb{R}$-automorph with compact factor space.

Proof. The discreteness of the group of integral matrices $\mathrm{Aut}_{\mathbb{Z}} f$ in the sequential topology of the Lie group $\mathrm{Aut}_{\mathbb{R}} f$ is clear.

In order to show the compactness of the right coset space of $\mathrm{Aut}_{\mathbb{R}}f$ over $\mathrm{Aut}_{\mathbb{Z}}f$ we must show the existence of a sequence (Z_j) of elements Z_j of $\mathrm{Aut}_{\mathbb{Z}}f$ and a subsequence

$$(Y_j) = (X_{k_j}) \qquad 1 \le k_1 < k_2 < \cdots$$

for any sequence (X_j) of elements X_j of $\mathrm{Aut}_{\mathbb{R}}f$ such that the sequence (Y_jZ_j) is convergent.

Because of the nondegeneracy of f, the matrices of $\mathrm{Aut}_{\mathbb{R}}f$ are of determinant ±1 so that the symmetric matrix $X_j^TX_j$ is positive of determinant 1. Hence there is a unimodular matrix U_j for which the matrix $W_j = U_j^TX_j^TX_jU_j$ is reduced of determinant 1. It follows from the reduction condition that the matrices W_j and hence also the matrices X_jU_j form a bounded sequence. Therefore, there is a subsequence (X_{k_j}) of (X_j) for which the sequence $(X_{k_j}U_{k_j})$ is convergent. Hence also the sequence

$$(U_{k_j}^T X_{k_j}^T A_f X_{k_j} U_{k_j}) = (U_{k_j}^T A_f U_{k_j})$$

is convergent. Since it consists of integral matrices, it follows the existence of a subsequence (1) of (X_j) such that

$$U_{k_j}^T A_f U_{k_j} = U_{k_1}^T A_f U_{k_1} = C$$

and hence

$$Z_j = U_{k_j} U_{k_1}^{-1} \in \mathrm{Aut}_{\mathbb{Z}}f$$

and the sequence

$$(Y_jZ_j) = (X_{k_j}U_{k_j})(U_{k_1}^{-1})$$

is convergent.

The finite presentability of $\mathrm{Aut}_f\mathbb{Z}$ follows from a theorem in the paper "On the units of orders" by Hans Zassenhaus (J. of Algebra, 1972).

BIBLIOGRAPHY

1. BODEWIG, EWALD: Matrix Calculus, 2nd revised and English ed. Amsterdam, North-Holland, 1959.

2. CASSELS, JOHN WILLIAM SCOTT: An Introduction to the Geometry of Numbers. Berlin, Springer, 1959.

3. DONALDSON, JOHN L.: "Minkowski Reduction of Integral Matrices," Math. Comput. 33 (1979), pp. 201-216.

4. GAUSS, C.F.: Disquisitiones Arithmeticae, English Translation by S. Clarke, New Haven, Yale University Press, 1965.

5. MINKOWSKI, HERMANN: "Diskontinuitätsbereich für Arithmetische Äquivalenz," Crelles J. Reine Angew. Math. 129 (1905), pp. 220-274; Werke II, pp. 53-100.

6. ZASSENHAUS, HANS: "On the Units of Order," J. Alg. 20 (1972), pp. 368-395.